Environmental Politics and Theory

Our current environmental crisis cannot be solved by technological innovation alone. The premise of this series is that the environmental challenges we face today are, at their root, political crises involving political values.

Growing public consciousness of the environmental crisis and its human and nonhuman impacts exemplified by the worldwide urgency and political activity associated with the consequences of climate change make it imperative to study and achieve a sustainable and socially just society.

The series collects, extends, and develops ideas from the burgeoning empirical and normative scholarship spanning many disciplines with a global perspective. It addresses the need for social change from the hegemonic, consumer capitalist society in order to realize environmental sustainability and social justice.

The series editor is Joel Jay Kassiola, professor of Political Science and dean of the College of Behavioral and Social Sciences at San Francisco State University.

China's Environmental Crisis: Domestic and Global Political Impacts and Responses
  Edited by Joel Jay Kassiola and Sujian Guo

Ecology and Revolution: Global Crisis and the Political Challenge
  By Carl Boggs

Carl Boggs' Previous Publications

Gramsci's Marxism

The Politics of Eurocommunism (coauthored)

The Impasse of European Communism

The Two Revolutions: Antonio Gramsci and the Dilemmas of Western Marxism

Social Movements and Political Power

Intellectuals and the Crisis of Modernity

The Socialist Tradition: From Crisis to Decline

The End of Politics: Corporate Power and Decline of the Public Sphere

A World in Chaos: Social Crisis and the Rise of Postmodern Cinema (coauthored)

Masters of War

Imperial Delusions: American Militarism and Endless War

The Hollywood War Machine: U.S. Militarism and Popular Culture (coauthored)

The Crimes of Empire

Empire vs. Democracy

Phantom Democracy: Corporate Interests and Political Power in America

Ecology and Revolution
Global Crisis and the Political Challenge

Carl Boggs

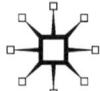

ECOLOGY AND REVOLUTION
Copyright © Carl Boggs, 2012.

All rights reserved.

First published in 2012 by
PALGRAVE MACMILLAN®
in the United States—a division of St. Martin's Press LLC,
175 Fifth Avenue, New York, NY 10010.

Where this book is distributed in the UK, Europe and the rest of the
World, this is by Palgrave Macmillan, a division of Macmillan Publishers
Limited, registered in England, company number 785998, of
Houndmills, Basingstoke, Hampshire RG21 6XS.

Palgrave Macmillan is the global academic imprint of the above
companies and has companies and representatives throughout the world.

Palgrave® and Macmillan® are registered trademarks in the United
States, the United Kingdom, Europe and other countries.

ISBN: 978-1-137-26403-9

Library of Congress Cataloging-in-Publication Data

Boggs, Carl.
 Ecology and revolution : global crisis and the political challenge /
Carl Boggs.
  p. cm. — (Environmental politics and theory)
 ISBN 978-1-137-26403-9 (hardback)
 1. Political ecology. 2. Environmentalism—Political aspects.
3. Environmentalism—Economic aspects. 4. Environmental
policy—Political aspects. 5. Environmental policy—Economic
aspects. 6. Global environmental change—Political aspects.
I. Title.
JA75.8.B64 2012
363.7—dc23   2012011143

A catalogue record of the book is available from the British Library.

Design by Integra Software Services

First edition: September 2012

10 9 8 7 6 5 4 3 2 1

Printed in the United States of America.

Contents

Foreword *Michael Parenti*	vii
Series Editor Preface *Joel Jay Kassiola*	xi
Preface	xvii
1 The Radical Imperative	1
2 The Global Crisis Worsens	23
3 The Political Impasse	57
4 Liberal Delusions	97
5 Struggle For an Ecological Politics	121
6 A Global Ecological Revolution?	153
Conclusion: A Green Politics?	191
Postscript: Ecology and Population	197
Notes	209
Index	223

Foreword

When the environmental movement burgeoned in the 1960s, it showed itself remarkably indifferent to questions of political influence and moneyed power. Most of the environmentalists of that day embraced New Age nostrums rather than radical political ideologies. It was their view that "pollution" was the core environmental problem, and that the task ahead was to clean up the rivers, lakes, and lands. This would be accomplished when we learned to produce and consume in ways that brought greater self-sufficiency and caused as little damage as possible to Mother Nature.

Some few of us, however, saw a troubling and even strangulating link between environmental devastation and the corporate pursuit of profit. I, for one, maintained that capitalism and ecology were on a collision course, and that people who wanted to save the environment would eventually have to confront the big corporations and the system that sustains the plutocracy. When I uttered such pronouncements in the classroom, several of my students who were active in the environmental movement were not too receptive. They had an agenda that did not include socialism. They pointed out that environmental degradation could be found in communist countries as well as capitalist ones, hence, the Reds had nothing to teach us on this subject.

Other New Left communitarians argued that political radicals should not try to hijack a social and cultural issue in order to rally support for their own quixotic goal of vanquishing capitalism, a crusade that was deemed not all that relevant to the problem of environmentalism. What was needed, they said, were improved living habits and more inventive and communitarian ways of production and consumption.

We political progressives would have done much better back in the sixties had we possessed a book such as the one by Carl Boggs, now in the reader's hands. In those days we felt that, while new ways of consuming and producing certainly were needed, there also was another whole politico-economic dimension to the environmental problem. Indeed, with the benefit of hindsight we now know that it is not merely a "problem" we face, it is a catastrophe of global magnitude. And it involves an all-consuming democratic political struggle of revolutionary dimensions, just as Boggs says.

The core process of global corporate capitalism is to transform living nature into commodities and commodities into dead capital—through the accumulation of profits. A central but mostly unspoken notion behind this ever-expanding investment process is the assumption that nature's reserves are inexhaustible. But running out of natural resources—such as oil—is not the central problem we face. Immense oil reserves are being discovered every year from the Arctic to Canada to Africa. The real danger comes when the planet's overall ecology can no longer sustain a livable dynamic. Well before we run out of oil and coal, we are likely to run out of fresh air, drinkable water, serviceable topsoil, livable coast lines free of increasingly destructive floods, and face various other unendurable weather aberrations connected with global warming. We now realize that the Earth's capacity to absorb heat from energy consumption is not limitless and the long-range effects can be horrendous. As Boggs puts it, we face a future that portends "outcomes too nightmarish to imagine."

And this future is not so far off. Much of it is already upon us. Ecological aberrations are growing in scope and magnitude at a speed considerably swifter than we had feared. It is not "our grandchildren" who will be affected (sweet little things that they might be). Disaster will not come upon us at "the end of this century" when over 90 percent of the people now alive will be dead. Disaster is happening right now in our own lives as the planet loses all its ice caps and frozen tundra, and as the gulf streams that create temporal zones slow down under the weight of massive ice meltdowns. The oceans are dying and if the oceans die, so do we, for they provide most of our oxygen. Indeed, there is a continual loss of oxygen in the air and water, and dangerous increases in ocean flood levels. One could go on. To quote Boggs, global warming "is likely to reach cataclysmic levels in just a few decades—a specter eliciting denials and stonewalling at the summits of power."

Boggs makes clear that were the vast private treasures of the super-rich more equitably distributed and rationally utilized—not for the purpose of still more capital accumulation but for housing, jobs, mass transit, energy alternatives, environmental protection, and human services—then a soundly based prosperity would be at hand, as measured not by private profit maximization but by human needs and environmental betterment (see Chapter 5, "A Global Ecological Revolution?"). As Boggs puts it:

> By stripping away the (false) relationship between corporate-defined "growth" and social well-being, an ecological outlook points toward heightened living standards while also dramatically reducing GDP and, with it, negative human footprints on the global habitat. The reality is that existing measures of "growth" conceal untold amounts of waste and destruction in the resources consumed by corporate superprofits, a lopsided emphasis on "private" over "collective" forms of consumption, a grossly inefficient energy system, a militarized economy, a meat-based agriculture and fast-food system, and a top-heavy finance capitalism.

What is called "growth" in today's transnational corporate world means ever-expanding, multi-trillion-dollar accumulations for the 1% along with social impoverishment and environmental devastation for the 99% of us. What is needed are not apolitical environmental half-measures but a revolutionary transformation of our politico-economic life, an empowerment of our democracy of a kind never seen before.

This book takes us into forbidden realities. By "forbidden" I mean the realities about wealth and politico-economic power that are seldom if ever honestly entertained in mainstream media or conventional politics. Prepare yourself for a commanding and revelatory investigation of politics, plutocracy, ideology, and ecology in all their challenging and disturbingly interrelated dimensions.

—Michael Parenti
www.michaelparenti.org

Series Editor Preface

Offering a Genuine Political Theory of the Environment: The Imperative for a Concrete Political Strategy for Ecological Revolution

... "agency" to be politically decisive, requires ideological articulation, organizational cohesion, and strategic direction.
—Carl Boggs, *Ecology and Revolution*, Chapter 4

As Editor, it is a privilege to introduce Carl Boggs' *Ecology and Revolution: Global Crisis and the Political Challenge,* the second volume to be published in the Palgrave Macmillan Series on "Environmental Politics and Theory." This Series explores the root social and value-based causes of the prevailing environmental crisis. Equally important to the mission of this Series is a focus upon the urgent need for societal transformation of our hegemonic, modern consumer society in order to achieve environmental sustainability and social justice. While the theme of the first volume published in the Series, *China's Environmental Crisis: Domestic and Global Impacts and Responses* (edited by Sujian Guo and myself), has been my intense recent interest, Boggs' subject matter in this volume—ecology and revolution, or the imperatives for implementing an alternative, desirable, sustainable society—has influenced my thinking, writing, and teaching ever since the present wave of environmentalism began in the 1970s with the publication of *The Limits to Growth* Report.[1]

Carl Boggs' trenchant and important discussion enriches this Series on a subject of vital significance to all living species inhabiting the planet. His analyses of environmental thought and prescriptions about the political requisites and prospects for global ecological revolution are insightful, persuasive, and, will hopefully, stimulate environmental scholars to turn attention at last to the inescapable problem of environmentally inspired societal transformation. Boggs' work should also appeal to policymakers and the general public, who must end their collective denial and come to terms with the unsustainability of capitalist consumer society evidenced by the current global crisis and its worsening (see Chapter 1). This raises the profound defining question of political theory and of our time: what social order should follow the present unsustainable one, and how can it be realized before ecological catastrophe strikes? Boggs' book focuses upon these important issues, and offers some solutions.

Over 20 years ago, after examining what was then exclusively biophysically based environmental limits-to-growth studies, I described the "absence-of-means-to-social transformation" fallacy, or more simply, the "absent of means" fallacy.[2] What I had in mind then—and now as well, since, alas, little has changed for the better—is illustrated in Boggs critiques of environmental thinkers (see Chapters 3 and 4). Hence, Boggs emphasis upon the political requirements of "durable organization, coherent ideology, and workable strategy" (Chapter 5), for a successful ecological revolution, is timely and significant. The message of *Ecology and Revolution* speaks directly to the widespread failure by the many students of the environmental crisis—both reformists and radicals—whose research clearly indicates the need to transform the thinking, values, and social structure of our neoliberal and consumerist culture, yet they do not go beyond the surface to address essential political question of *how* to implement the alternative ecologically, socially, and morally superior society.

This challenge of realizing one's theoretical visions is as old as Western political theory itself. Plato posed it in his foundational work, *The Republic*, when Glaucon asks Socrates about the ideal society: "Is it possible for this [ideal] regime to come into being and how is it ever possible?"[3] Socrates' response reverberates throughout the subsequent 2400 years of Western political thought. He considered Glaucon's question to be "the biggest and most difficult, the third wave;" the ancient Greeks traditionally thought of the third wave to hit the shore the biggest and fiercest wave.[4]

Therefore, to provide a devastating critique of globalized consumer society based on a litany of environmental threats without confronting the topic of the nature of an implied alternative sustainable society *and* how it would be realized (the "Glaucon question"), especially over the strong resistance of the beneficiaries and advocates of the status quo (like the corporate ruling elite) seems seriously, if not fatally, incomplete. Furthermore, even to suggest the outlines of any alleged alternative sustainable social order without addressing the politics of a transition or change strategy seems nothing less than utopian. The great merit of Boggs' volume is that he explicitly conducts this quintessential political inquiry into the multidimensional nature of the global crisis. His comprehensive and nuanced understanding goes beyond merely biophysical components to include economic, social, political, and military factors as well as societal transformation (Chapter 1).

While many researchers in the last 40 years of the environmental movement have engaged in the negative critique activity of this necessarily two-part project—beginning with the computer projections-based first report on *The Limits to Growth*—few have proposed detailed alternatives for an environmentally sustainable and socially just society; even fewer have conducted a genuine political analysis of its strategy and means of implementation. Boggs' distinctive discussion of this difficult but crucial task for societal transformation is vital for the future of our world. To my knowledge, sparse works exist that are as comprehensive in breadth, focusing upon the several dimensions of the current global crisis (Chapters 1 and 2); and no extant work that I am familiar with is as thorough in analyzing the need for the political means to radical social change, presented by Boggs as the organizational, ideological, and strategic elements of societal transformation (see Chapter 5).

In contradistinction to most of the limited nonpolitical works about the environmental crisis written by natural scientists, Boggs illuminates the links between ecology, politics, and societal transformation. Furthermore, he identifies where liberal environmental reformers and radical environmentalists seeking a social order superior to the unsustainable status quo fail to address the Platonic political issue of achieving one's alternative conception of society (Plato's "city in speech")[5] (see Chapters 3 and 4). Boggs' specifically political framework for societal transformation produces a counter-hegemonic agenda that has been sorely missing from the biophysically based environmental literature. Using this political framework, Boggs is able to make critical assessments of a wide variety of environmental thinkers: from Al Gore and Lester Brown, on the one hand, and Murray Bookchin, Deep Ecologists, and Left-Green Eco-Marxists, on the other (see Chapters 3 and 4). Boggs expresses this central theme, to my mind the most important contribution of his book, as follows:

To speak of "politics" means addressing the key requisites of fundamental change: a well-defined alternative ideology, dynamic leadership and organization, resources for mass mobilization, effective strategy, an orientation toward winning governmental power.

(Preface)

Following the examples of Plato and subsequent political theorists in the Western canon, Boggs emphasizes the politics of transforming our current neoliberal consumer society in order to avert some kind of disaster—ecological, economic, social, political, or military, which all overlap and are connected in the practical world. Realizing the severity and urgency of the current global situation, Boggs informs his readers early on that he has "written this book with the conviction that a radical *political* solution to the crisis is desperately needed if humanity is to escape a nightmarish fate" (Preface, emphasis in original).

Boggs' presentation of the global crisis is noteworthy for going beyond the standard description of the various natural threats facing humanity and the other living creatures, such as: climate change, resource depletion, pollution, deforestation, population pressures, agricultural challenges, etc. (Chapter 1). All of these have been worsened by intensified economic and cultural globalization since the 1990s. Yet, as Boggs perceptively notes, without a clear and comprehensive understanding of the need for an effective political strategy necessary to both defeat the ruling ideas and elite, and to conceptualize as well as implement an alternative sustainable social order, the current tragedy we are suffering under will continue to worsen. This profound global political tragedy is described by Boggs as follows:

> ... no political force now blocks the path of ever-expanding corporate power... Integrated corporate domination helps perpetuate a general mood of futility, disillusionment, and cynicism: if ecological crisis demands radical intervention, the prevailing ethos works in just the opposite direction—toward a settling in of normal economic and political routines.
>
> (Introduction)

Added to this pessimism regarding radical social change is the rejection and suppression of genuine politics and politically created societal transformation under the false guise of apolitical neoliberal economics by the dominant ideology of neoliberalism.[6]

Boggs' volume teaches us the urgent need to retrieve and legitimize the first importance of radical politics from depoliticization produced by corporate power and neoliberalism. The environmental crisis can be the catalyst for such a fundamental repoliticization (see Introduction). To attain such a revitalization of radical politics and the transformative political agenda of strategizing; building institutions; and developing ideologies; and, in short, politically preparing for societal transformation, Boggs turns to an updated and renewed Green Party and Social Movement (Chapter 5).[7]

Whether or not one agrees with Boggs' recommendations for achieving the global ecological revolution, merely considering and assessing it from a political revolutionary perspective will advance the profound social dialogue about the politics underlying the global crisis and our effective response to it. What is most important about Boggs' account of societal transformation is his emphasis upon the means to achieve it by envisioning how the theorized ideal is to be implemented: organization, leadership, and a strategy for success. Boggs admonishes those thinkers who understand the need for such societal transformation but fail to seek, or even conceive of an effective political strategy to achieve it. He explains why we have reached this grim state of political affairs with little time remaining before the necessary global ecological revolution must be accomplished: "No groups, movements, or parties have yet been able to forge a subversive political strategy viable enough to win critical masses of support" (Preface). Regarding the prospects of a successful transformation of the new worldwide neoliberal consumer society, I remind readers, as I do my students, that there has never been a revolution of a modern industrial society. Therefore, there is no precedent upon which to model the global ecological revolution, although it is desperately needed. The political challenges of achieving such a societal transformation today seem as insurmountable as the imperative to transform our unsustainable hegemonic social order. Is this the ultimate modern tragedy? To its great credit, Boggs' volume delineates the contributing factors to this potential tragedy and, thereby equips us to take informed action to address it.

Boggs' profound book with its imperative for "winning governmental power" exhorts us to start the difficult but unavoidable political work that begins with this recognition: *thinking about the environmental crisis must be politicized if we are to avoid disaster. Ecology and Revolution* asserts the necessity for a political means to the ecological revolution. I strongly urge the reader to reflect deeply upon the agenda so forcefully articulated by Boggs in this book, and to begin to carry it forward.

Joel Jay Kassiola

Preface

A book titled *Ecology and Revolution* and dedicated to an exploration of the global crisis has not been written with the understanding that all is right with the world as we know it. In this case, the focus is on a looming nightmare that will dominate the world landscape for years and decades to come: the global ecological crisis—a crisis unprecedented in both scope and urgency. The following pages explore the multiple dimensions of this epic predicament, with an eye toward framing the great environmental challenges of our time within a larger matrix of economic conditions, political forces, and international relations. In contrast to so much other work on this daunting topic, the present study looks primarily at its *political* side—or, more accurately, the dialectic linking ecology and politics, natural relations and social transformation. Devoting attention to politics in this fashion brings us to an investigation of historical struggles (their successes and failures, promises and frustrations) against the major centers of power responsible for ever-increasing levels of exploitation, waste, and destruction. To speak of "politics" means addressing the key requisites of fundamental change: a well-defined alternative ideology, dynamic leadership and organization, resources for mass mobilization, effective strategy, an orientation toward winning governmental power. Unfortunately, the legacy of ecological politics to date across the industrialized world has—with few exceptions—provided little cause for optimism, nowhere more so than in the contemporary United States. The leading superpower throughout the postwar years, the United States, owing to its global hegemony and unique contribution to the modern crisis, has special leverage and resources (not to mention ethical obligation) to help reverse the further slide toward barbarism—yet its power structure represents the main roadblock to such an outcome. And time is running perilously short: for the international scientific consensus warns that the planetary tipping point—the moment when human remedies will no longer suffice—is imminent. I have written this book with the conviction that a radical *political* solution to the crisis is desperately needed if humanity is to escape a nightmarish future.

The modern industrial assault on the natural habitat has intensified over the past two centuries and more, taking the form not only of climate change but an assortment of other (related) threats: depleted resources, vanishing arable land, polluted air, water, and soil, nuclear radiation, deforestation, urban deterioration, food shortages, and population pressures among them. Even if this assault were to end abruptly—highly unlikely under present circumstances—the harmful consequences of long-term buildup would persist for hundreds of years. In fact a reversal is nowhere in sight; on the contrary, the ruling corporate, governmental, and military interests (assisted by a propagandistic media) work indefatigably to *block* far-reaching change despite much talk at the summits of power about environmental reform, renewable energy, and "sustainable" growth. Given existing power configurations in the United States and other leading industrialized nations, the prospects could hardly be different as vast resources at the disposal of transnational business empires, the banking plutocracy, agribusiness interests, permanent war economy, and corporate media are thrown against mostly dispersed, far weaker local communities, workers, and popular movements. Insofar as this power structure is non-reformable from the standpoint of effectively confronting the crisis, the only choice for humanity as a whole is a global ecological revolution. The following pages address the different, and increasingly problematic, facets of this political imperative.

A planet hurtling toward escalating chaos and destruction is the logical result of a world capitalist system driven to accumulate unlimited wealth, resources, and power. In its hell-bent efforts to commodify and dominate every region of the globe, the corporate-growth machine recognizes few constraints on its *modus operandi;* the system is inherently and irrevocably anti-ecological, the mortal enemy of nature. There can be no egalitarian or democratic, much less *sustainable,* capitalism whatever the claims of liberal reformers in support of a "green" market economy. In the guise of freedom, democracy, and progress, the world system engulfs virtually every realm of human (and nonhuman) life, destroying ecosystems while resisting even minimum levels of popular accountability. This logic applies especially to the United States, reputedly a model democracy but ruled by a narrowing stratum of oligarchs aligned with supercorporations, financial institutions, the military, and the media-entertainment complex. Those interests are so tightly integrated, so reinforced by their media venues, lobbies, think tanks, and action committees, that American election and legislative activity has come to matter less with each passing year; access to all but the super-wealthy elites and their tireless propagandists is effectively closed. Such trends coincide with the general downward trajectory of a society in the grips of worsening poverty, massive unemployment, eroding public services, a failing infrastructure, obscene gulf between rich and poor, and environmental ruin. Observing this extraordinary waste and destruction, but soothed by increasing superprofits, the ruling stratum prefers obstruction over change, reaction over reform, mythical bromides like "free markets" over creative problem solving. Perhaps this is to be expected for a stratum that, worldwide, constitutes less than one percent of the population but controls nearly 50 percent of total wealth.

Insofar as an increasingly oligarchic, global, and solidified power structure cannot be reformed (much less "greened") toward sustainability from within its own boundaries and rules, options for solving the ecological crisis—and kindred challenges—are inescapably narrowed. It follows that radical alternatives will have to find their way onto the political agenda, and soon. This is a particularly imposing task since opposition to the corporate-growth madness has so far been confined to either comforting liberal refinements or localized grassroots movements—again perhaps nowhere more so than in the United States. No groups, movements, or parties have yet been able to forge a subversive political strategy viable enough to win critical masses of support. The task is rendered even more difficult once the futility of earlier change models is recognized: social democracy as institutionalized loyal opposition, Leninism as a formula behind authoritarian party-states, anarchism or syndicalism as a recipe for political isolation and impotence. The Green synthesis that surfaced in Europe three decades ago did offer the renewed hope of an "antiparty party" uniting electoral and movement struggles, but it too quickly followed the social democratic path of deradicalization. A program of even the most worthwhile liberal reforms, as I emphasize throughout this book, will be entirely too inadequate and time consuming to reverse the global crisis. The Greens, at least, did lay the ideological foundations of a post-liberal, post-Marxist transformative politics needed to finally evict a deeply entrenched ruling elite. The vision of a rejuvenated, globalized, and more combative Green politics constitutes the recurrent Leitmotif of this book.

The long historical struggle for a more egalitarian, democratic, and livable planet demands now, more than ever, a thoroughgoing rupture with the past, a shift toward new modes of being—not only fundamentally altered forms of production and consumption but altered lifestyles, beliefs, habits, identities, and above all *politics*. It also requires a transcendence of failed legacies of the past, whether liberalism, nationalism, anarchism, or social democracy, as we navigate toward a global ecological revolution—though positive elements of those legacies can surely be appropriated within a new framework. Most conventional theories and strategies were in fact long ago made obsolete by drastically changing material, institutional, social, and above all ecological, conditions. Beyond mere *choice*, we have entered the rapidly changing global realm of moral and political imperatives.

My work on this book has been greatly sustained and immeasurably enriched by the enduring personal and intellectual support of Laurie Nalepa.

Carl Boggs
Los Angeles
April, 2012

Chapter 1

The Radical Imperative

The global ecological crisis, rapidly approaching a point of no return, threatens planetary survival at a time when countervailing forces have so far been unable to resist, much less overturn, the powerfully destructive forces at work. The crisis intersects with, and reinforces, virtually every challenge human beings face, from chaotic weather patterns to the depletion of natural resources, the spread of poverty worldwide, the erosion of public infrastructure, impending agricultural disasters, and the likelihood of widening military conflict. The underlying cause is globalized corporate power, now hell-bent on commodifying and dominating every community, workplace, cultural space, and natural habitat on the planet. Fragile ecosystems face increasing threats while the ruling interests and their propagandists justify the perpetual growth machine as vital to social progress and material prosperity. Although warfare between and within nations has long appeared as a normal state of global affairs, perhaps the most devastating war is the one being waged by humans against nature. Opposition to the looming global catastrophe can make little headway until it breaks the tightening hold of a power structure that valorizes nothing so much as the limitless pursuit of wealth, resources, and hegemony on a world scale—and the time is growing short. The argument set forth in the following pages is that the modern crisis demands an uncompromising radical politics oriented to qualitatively new modes of production and consumption across the globe.

The world is currently experiencing the biggest and deepest upheavals that ecosystems have endured for millennia—a product of sustained industrial development through more than two centuries. Thanks to independent scientific work from dozens of countries, questions about the underlying *sources* of the ecological crisis have been definitively answered: the major villain is human activity in the form of greenhouse emissions from fossil fuels, augmented by waste and destruction endemic to the world capitalist system. Even if this merciless assault were to end abruptly—not likely, given the scope and momentum of the corporate growth apparatus—vast environmental changes set in motion long ago are already sufficient to melt all planetary ice in a matter of decades, with irreversible consequences. As of this writing (late 2011), attempts to reverse the global crisis have been so tepid, so half-hearted, so begrudging as to make little difference to immediate global prospects; corporate and government business-as-usual goes on, above all in the leading industrial nations that bear most responsibility for the crisis. According to well-researched scientific reports sponsored by the United Nations and major universities, melting glaciers could boost world sea levels by three feet in just a few decades, wreaking havoc on coastal areas, dozens of cities, the world economy, weather patterns, food production, public health, and, most probably, the prospects for democratic governance and international peace. Even more disturbing, if most or all ice on earth does eventually melt, sea levels could rise well *beyond* three feet, with outcomes too nightmarish to imagine. The first ten years of the twenty-first century were in fact the hottest on record, reflecting a phase of climate change already linked to extended droughts, floods, perilous storms, and vector-borne diseases, problems likely aggravated by ongoing rainforest devastation.

The planet has moved from 280 ppm (parts per million) greenhouse emissions in the 1880s, as the industrial revolution was taking off, to a staggering 390 ppm in 2010—an upward trajectory that, unfortunately, shows no indication of abating in the midst of full-scale corporate globalization and widespread political paralysis. A National Oceanic and Atmospheric Administration (NOAA) 2011 report found that global-warming related pollution (carbon, methane, nitrous oxide, CFC 11, and CFC 12) rose by 1.5 percent from 2009 to 2010, a far steeper rise than had been expected by the International Panel on Climate Change (IPCC), sponsored by the United Nations, and many independent scientists.[1] According to the NOAA Annual Greenhouse Gas Index, moreover, total emissions had increased by 29 percent since 1990. The world pumped 564 million *additional* tons of carbon alone into the atmosphere from 2009 to 2010, with the United States and China far outdistancing all other national sources—well beyond the worst-case scenario outlined by the IPCC. The Carbon Dioxide Information Analysis Center, run by the U.S. government, announced that greenhouse gas emissions jumped by 6 percent from 2009 to

2010, far surpassing earlier projections.² All reports identified a small group of usual suspects: fossil fuels, coal-fired plants, petrochemical and related industries, and commercialized agriculture.³ They further noted that, despite noble promises to solve the problem of climate change, little effective political action had been undertaken.

Such terrifying scenarios, reforms or no reforms, mean that the fearsome tipping point—a global threshold of no return—is probably not too distant. As the crisis escalates, moreover, ecological ramifications will be impossible to separate from parallel disasters in the economy, politics, social life, and international relations. Even if all nations were to urgently address the challenges at hand, humanity would still need hundreds if not thousands of years to return the oceans, waterways, air, soil, forests, and planetary atmosphere back to healthy, livable, and sustainable preindustrial levels. But as essentially casual responses and timid solutions at Rio de Janeiro (1992) and Copenhagen (2009) have shown, such readiness for action by national and world leaders is unfortunately nowhere in sight. As of this writing (late 2011), the rather moderate agenda set for the UN-sponsored Climate Change Conference at Durban, South Africa, did not inspire much cause for optimism.⁴

The crisis is sure to be felt most immediately in the realm of agriculture and food production, threatened by hotter temperatures, shrinking arable land, long-term droughts, massive flooding, falling water tables, soil erosion, and resource-wasteful animal farming. The question of how, with intensified global warming, the world will be able to feed a population of at least nine billion by the year 2050 has yet to be addressed by any major government or political organization. Food shortages will become extreme, fueling such disasters as famines, epidemics, and social dislocations, with tropical areas most severely hit. World grain production, having peaked in the early 1980s, is declining at the very moment global demand is rising sharply. Most countries can anticipate a 5 percent or more drop in agricultural output within the next decade, setting in motion a worldwide desperate search for imported foodstuffs—where in fact such imports are available. More than three decades of intensifying global warming has meant declining worldwide yields of corn (by 5.5 percent) and wheat (by 3.8 percent), a trend sure to bring more shortages, increased hunger, and higher prices.⁵ As Lester Brown points out, the world's irrigated land area tripled between 1950 and 2000 but has expanded little since then.⁶ Water shortages, already severe, will further aggravate both food and health challenges. Food riots erupted in scattered parts of the world during 2008 and 2009, anticipating a future norm when shortages mount and prices skyrocket. (Worldwide food prices had already increased by 80 percent between 2005 and 2008, a hint of deeper trends at work.) These challenges mount with the emergence of food as a major commodity within the global casino economy, as Wall Street banks and investment firms seek billions in profits while scarcity and hunger afflict tens of millions of people. This situation is exacerbated by the growth of meat-based agriculture and the fast-food economy, especially in the industrialized societies, as some 80 percent of grains are currently fed to livestock while meat production as a whole devours three times more soil, water, and fossil fuels than cost-effective plant-based food systems. (In the United States nearly 60 percent of all farmland is devoted to beef production, which also consumes roughly *half* of all water for human purposes.) The expansion of large-scale, centralized agribusiness, a source of vast waste and pollution—when added to the repercussions of global warming—could itself push the crisis far beyond what has been projected.

Imminent water scarcity is a salient case in point. By 2010 several multinational corporations were vying to turn water resources into a profit-making commodity, overturning nearly 2000 years of history during which water had been protected within the public domain. Water shortages are destined to grow as capitalist markets—with global demand outstripping supply—drive up prices, denying millions of people access to a resource most needed for agriculture, transportation, industry, the public infrastructure, indeed every realm of human activity. With global water consumption doubling roughly every 20 years, climate patterns becoming more chaotic, aquifers being depleted, and pollution contaminating more lakes, rivers, and oceans, privatization will allow controlling business and banking interests to fix prices at whatever the market can tolerate, further accelerating the path toward ecological ruin.[7]

With the crisis intensified by resource shortages—first in energy but also in timber, water, scarce metals, and even land—the world can expect sharpening national conflict accompanied by heightened military spending, a flourishing arms trade, spreading civil strife, and new outbreaks of warfare. U.S. geopolitical designs in the Middle East, a region laden with petroleum and other resources, could be read as an opening chapter in the looming era of resource wars. Michael Klare argues that the end of the Cold War has brought natural resource priorities to the forefront of U.S. foreign policy and military planning.[8] As is well known, the United States today spends as much on military force and worldwide deployments as all other nations combined, with bases scattered across more than 100 countries and plans afoot to modernize nuclear weapons and militarize outer space. These conditions, naturally, coincide with moves by rising powers (China, India, Brazil, Mexico, Venezuela, etc.) to strengthen their own arsenals as they seek leverage (and perhaps deterrence) in a conflict-ridden world. While such countries look to expand their military and possibly even nuclear capabilities, often at the expense of social needs, terrorist groups could thrive in a context of blowback as the most powerful nations elevate their arms spending and geopolitical ambitions. International treaties—for example, prohibitions against military aggression, the Nuclear Non-Proliferation Treaty (NPT), the ban on space militarization, International Criminal Court—could be reduced to scraps of paper. In this Hobbesian atmosphere, trends toward militarism, economic competition, political chaos, and ecological deterioration could push the global crisis to new levels.

If a politics of radical change needed to meet the global challenge nowadays seems remote or utopian, that is to be expected: alternatives to the world capitalist system are presently weak and fragmented, and what enclaves of opposition exist in the industrialized countries typically lack political leverage, cohesion, and durability, not to mention critical mass. Some liberal and progressive forces have mobilized enough resources to win limited reforms, but few have the capacity to disrupt business-as-usual and none pose a serious threat to the power structure. No anti-system movements or parties exist with much scope or permanence, and that includes a multiplicity of "green" tendencies in Europe and elsewhere. Political impasse is of course hardly unique to environmental politics: long-standing human goals such as disarmament, containing nuclear proliferation, abolishing poverty, eradicating serious diseases, and extending democratic politics, for example, are no closer to realization today than they were several decades ago. In the aftermath of an earlier worldwide collapse of socialist politics and the more recent spread of right-wing ideology in the United States and Europe, oppositional forces have stalled or retreated, with anticapitalist initiatives effectively countered within the political and media establishments. The tragedy is that while radical alternatives end up thoroughly discredited, they are the *sine qua non* of averting planetary catastrophe. At the same time, history does show that even limited, tenuous popular victories can set in motion far-reaching change, often in the face of seemingly impregnable power structures—as in the cases of South Africa, the former Soviet Union, and Eastern Europe. The Zeitgeist of political retreat could be more difficult to defend as fissures and cracks in an outwardly sturdy corporate-military structure widen and global capitalism reaps more and more of its own bitter harvest.

World events since the early 1980s have fueled a steady rightward shift: the Reagan presidency, Soviet collapse and end of the Cold War, U.S. military interventions in the Middle East, business and financial deregulations, 9/11 and its aftermath, intensified globalization, and a tightening corporate stranglehold over the mass media. Both Chris Hedges and Tariq Ali recently observed that American liberalism now stands corrupted and eviscerated, perhaps best illuminated by the steady retreat of the Barack Obama presidency; no political force now blocks the path of ever-expanding corporate power.[9] Meanwhile, U.S. global power retains its hegemonic aspirations despite widespread talk of imperial decline, helped along by ideological justifications stemming from the post-9/11 war on terrorism. Despite contradictions endemic to the world system, the power of capital (backed by military force and transnational structures like the International Monetary Fund and the World Bank) had by the early years of the twenty-first century penetrated more extensively across the planet. Integrated corporate domination helps perpetuate a general mood of futility, disillusionment, and cynicism: if ecological crisis demands radical intervention, the prevailing ethos works in just the opposite direction—toward a settling in of normal economic and political routines. With socialism historically discredited as the ideological matrix of oppositional politics, moreover, sources of radical change now become more difficult to locate than in the past. Liberalism, as mentioned, is far too embedded in the power structure to fill the ideological void.

Despite the familiar requiem for socialism, it seems useful nonetheless to resurrect Rosa Luxemburg's famous dictum—"socialism or barbarism"—though now in a rather different guise. The same imperative was appropriated several decades ago by European radical theorist Cornelius Castoriadis in his early post–World War II writings.[10] Both Luxemburg and Castoriadis would presumably agree that the grim but unavoidable choice today, put starkly, is between radical change and planetary collapse. The corporate-driven world system follows an indelible pattern of perpetual growth, maximum deregulation, and endless resource consumption—agendas bereft of ethical precepts and fully at odds with sustainable development. As the leading economic and military power—and the main source of the environmental crisis—the United States is uniquely positioned to help reverse the crisis: if an ecological politics fails to gain traction in American society, then worldwide hopes are likely crushed. Yet for

Washington elites looming disaster nowadays elicits mostly formulaic responses mixed with outright denial, as the power structure moves to relegitimate itself on an archaic foundation of neoclassical economics and "free-market" bromides taken from nineteenth-century theory.[11] As oligopolistic power expands, the ideological turn toward free enterprise, privatization, and extreme individualism coexists with real-life authoritarian trends. The rebirth of traditional liberalism (read: modern conservatism), a product of early capitalism, more accurately signals a profound turn toward ideological escapism rooted in the fictions of a self-regulating economy, rugged self-reliance, and small-town America.

This contention might seem puzzling at a time when ideological volatility and social conflict appear on the upswing, with vibrant grassroots movements said to oppose "big government," establishment politicians facing harsh attacks, and congressional debates filled with angry name-calling. Electoral battles have grown more heated, nasty, and bitterly fought out, well-funded lobbies are increasingly aggressive, ideologically driven think tanks and foundations thrive, and the mass media is saturated with vituperative exchanges over such issues as immigration, gay marriage, health care, taxes, and public spending. In this context, environmental organizations proliferate around centers of governmental activity, pushing modest reforms while supposedly dedicated to saving the earth and "sustainable development." Critics like Thomas Frank and Robert Kennedy, Jr., have shown that many such groups have been cynically bought off and neutralized by the monied intervention of corporate lobbies.[12] Some grassroots oppositional movements have gained headway since the early 1990s, including "anti-globalization" protests that first surfaced in Seattle, immigrants-rights actions, and large-scale peace demonstrations before and just after the 2003 U.S. invasion of Iraq. Smaller movements, based in local communities and college campuses, appear from time to time and often manage to thrive. Popular mobilizations fought right-wing efforts to destroy public labor unions and their bargaining rights, as in Wisconsin and elsewhere during early 2011, with popular agitation reaching wider sectors of society, tapping into the mood of anger and resentment without, however, evolving into a broader political formation. Much the same could be said of the militant Wall Street Occupied movement that spread from New York City to dozens of American cities and abroad in late 2011. In fact no oppositional tendency has yet adopted an efficacious *political* strategy based in durable organization and methods adequate to sustaining a transformative presence mainly outside the established parties. What typically occurs within the dominant public sphere—elections, legislation, interest group lobbying, media debates—sidesteps much-needed debates over corporate, financial, and military power. Historical sources of anti-system opposition—anarchism, socialism, Communism, the new social movements, Green politics—have been marginalized or expunged from American political culture, which turns more ideologically parochial and narrow with each passing year. Nowadays even progressives have distanced themselves from those venerable legacies, convinced they could never generate transformative politics.

As American society accelerates its slide into a rationally administered order, consolidated elite power coexists with high levels of mass alienation, social fragmentation, and revulsion against "government intervention" (aside from the military, intelligence, and law enforcement). The result is a perpetual shrinkage of politics as citizen participation, public discourse, and social governance erode. Sources of general disempowerment are scarcely difficult to locate: concentration of corporate and financial power, bureaucratization, economic globalization, lobby stranglehold over elections and the legislative process, workplace authoritarianism, and the narcotizing influence of commercialized media culture. As the system takes on heightened oligarchic and global features, decision making further gravitates toward a tiny governing clique along lines of what C. Wright Mills identified several decades ago in his classic, *The Power Elite*.[13] The postwar expansion of U.S. global power, augmented by antidemocratic features of the war economy and security-state, has drained vast human, technological, and material resources away from public goods and services. U.S. imperial ambitions depend on sustained economic growth, exorbitant levels of military spending, worldwide intelligence and surveillance capabilities, and global armed forces deployments, all legitimated by superpatriotism that lies at the core of American exceptionalism. Empire becomes intelligible to mass publics insofar as it is organically linked to a common, routine "way of life" embedded in daily existence.[14] Despite episodic tremors that might disturb or threaten the political establishment, the reality of Empire in all its institutional and material resources remains beyond political debate, its accountability to citizens purely formal. As ideologically driven American leaders steer the nation—possibly the world—toward a new phase of "barbarism," political opposition is routinely channeled into safe venues; oligarchical power trumps social change at every turn. This is less a matter of specific institutional arrangements like the two-party system, winner-take-all electoral districts, and gerrymandered representation than of the systemic corruption and diminution of politics itself. Since the ancient Greeks, politics has been understood by philosophers as central to human existence, the foundation of public governance, community life, citizen participation, and creative statecraft. Theorists as diverse as Aristotle, Rousseau, Machiavelli, Marx, and Lenin viewed political discourse and action as indispensable to collective identities, the public interest, and social change. It was Aristotle who first embraced politics as a central arena of public interaction, vital to the sustenance of social life.[15] Little of this remains in the United States today, despite all the ritual celebrations of freedom, democracy, and citizenship.

With the widening scope of elite power, American politics allows little space for badly needed social reforms, much less radical change, as corporations, banks, and their lobbies set limits to policy choices and decisions within the two-party stranglehold. Money effectively determines most electoral outcomes, corporate and military lobbies shape congressional legislation, and the same interests colonize such regulatory bodies as the Environmental Protection Agency, Food and Drug Agency, Tennessee Valley Authority, Bureau of Mines and Minerals, and Federal Communications Commission. With rare exceptions, senators and representatives depend on oil, coal, and natural gas largesse, as well as insurance, pharmaceutical, and banking money, for reelection. In Congress, legislation not palatable to business interests is usually defeated: growth and profits easily prevail over environmental and other social priorities. With planetary survival in question, multinational giants like ExxonMobil, Microsoft, AT&T, Citibank, and British Petroleum—villain of the 2010 Gulf oil disaster—operate with relative impunity, subject to loosened or poorly enforced regulations, laws, and ethical constraints. In shameless protection of these interests, American politicians wind up as scientific and environmental "skeptics," with Republicans especially deriding climate change as a liberal fairy tale, at a time when weather-related disasters—droughts, floods, tornados, hurricanes, et cetera—were becoming the most widespread, fearsome, and costly in U.S. history. Many academics as well, often reliant on corporate largesse for their "research," have joined the denial frenzy. One example: Professor Roy Spencer, a scientist at the University of Alabama, has concluded from his work on climate change that, while heat continues to build up in the atmosphere, it escapes just as quickly and thus poses no threat.[16] In an illuminating investigative report, Naomi Klein, writing in the *Nation*, describes at length the global-warming denial machinations at the right-wing Heartland Institute, which held its sixth International Conference on

Climate Change in late 2011. Conference delegates were motivated by one all-consuming interest: to reject the scientific consensus that human activity is responsible for warming the planet. The deniers were highly imaginative in their concerns, some believing that climate change is a plot to steal American freedoms and democracy, others convinced that it is simply "a stalking horse for National Socialism," yet others taking the view that it is President Obama's scheme to deliver the United States over to some form of Communism. What apparently connected everyone at the conference, however, was deep fear that political efforts to confront the ecological crisis will be threatening to capitalism, to what delegates understood to be "the American way of life." It is this innately anticapitalist thrust of ecological politics, Klein aptly notes, that American ultraconservatives seem to grasp in a way that generally eludes liberals and progressives. It is this awareness, more than empirical data regarding climate change, that lies at the heart of warming denial.[17]

In the midst of a tenacious economic crisis, moreover, both politicians and mass publics remain fixated on material concerns, as issues like taxation, fiscal deficits, and public debt rise to the surface. One result of this fixation is that American public opinion in response to the threat of global warming has shifted remarkably in just a few years: surveys reveal a broad turning away from preoccupation with climate change, no doubt in response to both the economic downturn—reinforced by the capacity of Republicans to steer public debates toward matters of "austerity"—and the aforementioned growing culture of denial fueled by big-business interests.

Corporate power similarly works against prospects for *global* intervention, as shown by depressing results at the Copenhagen world summit in December 2009. The environmental gathering of 190 nations ended without binding agreements to restrict carbon emissions, the most influential delegates preferring voluntary accords without specific national and international targets—and no legal treaties to force governments and corporations away from their standard *modus operandi*. Some $30 billion was earmarked to assist poor countries in defending against the worst effects of climate change, at best a temporary palliative. After 15 climate summits and 20 years of scientific warnings about imminent ecological disaster, with irrevocable evidence that the planetary tipping point is near, the ethos at Copenhagen—exemplified by the U.S. delegation—was astonishingly relaxed, devoid of any sense of urgency or recognition that "market" devices are utterly inadequate to so much as mitigate global warming.

In fact Copenhagen, like the 1992 Rio failure, did nothing more than reaffirm commitment to the very neoliberal model that must be jettisoned if the modern crisis is to be solved.

This is the place to emphasize that the capitalist obsession with "economic growth," central to neoliberal ideology as well as the vast majority of environmental movements fixated on "limits to growth" or "no growth," is riddled with outlandish fictions, distortions, and myths. A qualitatively new (nondestructive, minimally wasteful, sustainable) model of human progress inevitably poses the question of precisely how growth is measured and interpreted. As it stands, the "growth" mania represents a cover for uneven development, which happens to coincide with further exploitation, social inequality, and environmental ruin. It is offered as something of an economic (and political) fix like the election campaign rhetoric of fiscal stability, "peace," and "family values." On the other side, calls for drastically reduced growth or "no growth" by environmentalists not only undercut popular appeals behind ecological politics as they seem to urge a regimen of harsh austerity and hardship, but replicate the very myths of neoliberal discourse. In reality the present "growth" system reproduces immense amounts of waste, destruction, inequality, and imbalance, all hidden beneath a fetishism of "free markets," "free trade," and aggregate measures of gross domestic product (GDP). Viewed thusly, the "growth" regimen bears little relationship to the actual quality of life that general populations experience: standard economic indicators of aggregate and quantitative "growth" fail to measure what matters most for people's everyday lives, as they are derived from outmoded calculations of the *overall* production and distribution of goods while failing to measure that specific *content* of those goods. By stripping away the (false) relationship between corporate-defined "growth" and general social well-being, an ecological outlook calls attention to heightened living standards that can be achieved by dramatically *reducing* GDP and, with it, shifting toward a vastly more egalitarian form of consumption, greater attention to the generation of public goods and services, demilitarization leading to social conversion, and a greatly lessened carbon footprint on the global habitat. Familiar measures of "growth" conceal untold amounts of waste and destruction in the resources devoured by corporate superprofits, a lopsided emphasis on "private" over "collective" modes of consumption, a grossly dysfunctional energy system, a debilitating permanent war economy, a meat-based agriculture and fast-food system, and a top-heavy, parasitic finance capitalism. While generating increased profits and wealth for a small elite, conventional "growth" agendas promise nothing so much as material, social, and environmental impoverishment on a world scale.

It follows that even modest reductions in systemic waste and destruction could mean dramatic cuts in GDP along with markedly *increased* living standards for the general population. More extended, deeper changes would of course allow for giant steps toward economic rationality and environmental sustainability. From this standpoint, GDP turns out to be a mirage, a misleading indicator of wealth, well-being, and social development. In settings where resources are the most unevenly distributed, as in American society, GDP levels (nearly $15 trillion for the United States in 2010) are even more distorted. The corporate-state obsession with "growth" therefore ends up significant primarily as an elite maneuver for superprofits and legitimation. At the same time, the deep-ecology choice between "growth" and "no growth" is false as it takes official GDP measures uncritically, at face value. Human needs can be much better satisfied with qualitatively fewer resources (perhaps half of existing levels) while extensively reducing destructive societal footprints on the natural habitat. (References throughout this book to "limits to growth," often in passing, as part of an ecological politics should be understood from this standpoint.)

Ecological radicalism depends above all on a vigorous retrieval of politics—a theme more fully explored in Chapter 5. As escalating anger over unemployment, poverty, deteriorating working conditions, eroding health care, and political disempowerment widens, resultant ideological shifts can turn rightward or, where leftward, end up devoid of the political strategy needed to win governmental power. Movements in the United States since the 1960s have most often remained pre-political or evolved into interest group liberalism, consistent with the American tradition of quick, instrumental results within a business-oriented culture. The antiwar, immigrants-rights, and global justice movements of the past two decades have largely followed a pre-political (or indeed antipolitical) trajectory. Eric Hobsbawm long ago observed that modern history is replete with pre-political or "primitive" types of rebellion: social banditry, millenarianism, urban riots, Luddite insurrections, and so forth.[18] While some of these forms were assimilated over time into "modern" party structures (social democratic, Communist, Green), others succumbed to localism and isolation before eventually departing the historical scene. To wrest power from elites owning vast power and wealth and ready to defend their class privilege by any method available, social movements have little choice but to pursue high levels of political articulation—that is, to become *modern* in both the Marxian (transformative) and Weberian (power-wielding) sense of historical efficacy.[19]

Modern politics, of course, can take many paths—military dictatorships, fascism, Communism, Keynesian state-capitalism, and social democracy among them. American society has taken the course of militarized state-capitalism since World War II, its ruling elites having secured maximum degrees of legitimacy and flexibility. Across the globe for the past century or more, political opposition has been overwhelmingly Marxist, or more accurately socialist, in its ideological self-conception. Three generic socialist "models" gained currency between the 1890s and 1920s: European social democracy, Leninism, and popular radicalism in the arc of anarchosyndicalism and council communism. The first two managed to win governmental power under entirely different conditions; the third ultimately disintegrated from its own isolation and futility, from repression at the hands of a superior organized force, or from assimilation into overarching party and union structures. No socialist politics, however, ever followed the Marxist theory of a proletarian-based anticapitalist revolution, and none (even where officially embracing socialism) could sustain antisystem politics for long. Social democracy first gained a foothold in Germany, spreading rapidly across Europe and winning power in several countries through expanded trade unionism and electoral gains. But initial hopes for a gradual, peaceful transition to socialism quickly gave way to welfare-state capitalism in such countries as Sweden, Denmark, Holland, France, and England. From the 1930s to the present, these parties evolved into institutionalized fixtures of state-capitalism, serving as a loyal opposition with diminishing transformative ambitions. Leninism served as the edifice of vanguard parties for the great twentieth-century revolutions, starting in 1917 Russia. Communist successes—in China, Vietnam, Yugoslavia, and Cuba as well as Russia—were simultaneously nationalist and multi-class, fueled by a generalized mass struggle against foreign domination and goals of national independence and economic modernization.[20] In contrast to social democracy, Communist triumphs came in preindustrial nations subjected to imperial domination and, in some cases, military occupation. In neither situation was the industrial working class a decisive force behind mass mobilization—nor indeed could it have been, given its relatively small numbers. If radical alternatives to social democracy and Leninism were historically condemned to political impotence, they did however survive as social and intellectual forces across time, resurfacing in the contexts of neo-Marxism, anarchism, the 1960s new left, European extraparliamentary opposition, and new social movements of the 1970s and beyond. As before, this new radicalism only rarely escaped its political impotence; local, dispersed, and pre-political, it never generated durable (organizational and ideological) alternatives to entrenched power structures. In the case of neo-Marxism (or "Western" Marxism), its destiny was appropriately characterized by Russell Jacoby as a "dialectic of defeat," though Gramsci's revolutionary politics stands as a noteworthy exception.[21]

Despite an episodic upsurge of popular movements and heated electoral contests, the broad American public has grown increasingly conservative, visible in a strict bipartisan consensus on an aggressive foreign policy, the rightward shift among both Republicans and Democrats, the spread of hostile attitudes toward government and "politics," the collapse of liberalism, and the rise in popularity of the Tea Party. Emotionally charged debates over such issues as gay marriage, immigration, and fiscal deficits—the stuff of media attention—have had relatively little bearing on the general contours of economic, foreign, and military policy. In the United States, to be sure, a postwar history of movement insurgency—Civil Rights, antiwar, feminist, ecology, gay rights, immigrants rights—has reshaped public life in areas of social legislation, legal reforms, popular consciousness, and working conditions. While earlier movement gains in the face of political obstacles and ideological backlash have been preserved to varying degrees, such gains never addressed class and power relations: the workings of corporate power, the war economy, the security-state, and geopolitical ambitions remain undisturbed, now as before. During the past three decades political debates have been overwhelmingly framed by conservative interests in business, government, the military, and media. The decisive power of such interests has been especially visible in the media, a point best articulated by David Brock in *The Republican Noise Machine*.[22]

With the socialist tradition dormant and popular movements choosing between political isolation and moderate reformism, radical energies resurfaced in the matrix of Green parties, first in West Germany during the early 1980s, then across Europe and elsewhere—raising the promise of a novel "antiparty" politics. Bringing movement agendas and styles into the realm of electoral politics, the Greens arrived at a political strategy to give dispersed movements the leverage of institutional power. Building on a merger of environmental, feminist, antiwar, and community groups, the Greens sought new modes of political articulation consistent with efforts to win governmental leverage. Radical discourses were injected into the public sphere, at times with exuberant flair and sense of experimentation.[23] From the outset, however, the West German Greens were firmly divided between radical (Fundi) and moderate (Realo) wings, the latter preferring electoral and governmental alliances with Social Democrats, the former more closely aligned with local movements. Over time, in Germany and elsewhere, the Realos prevailed: the seductions of power, the pull of institutionalization, and the moderating force of electoral politics were too much to resist.[24] Green literature retained obligatory nods to radical change, but everyday politics carried the parties inexorably toward social democratic reformism. By the mid-1990s no Green organization in Europe held to its original radical identity, having morphed into an institutionalized loyal opposition within less than a decade. In the United States, where after repeated campaigns the Green party achieved sporadic, local electoral victories, it never had radical pretensions, its outlook a mélange of Realo and new-age ideologies. Socialist ideas that often surfaced within the European Greens, giving rise to ecosocialist tendencies, were scarcely a factor in the United States, as party membership was drawn overwhelmingly from a youthful (mostly white) counterculture.

A vexing problem remains: in the absence of an anti-system political identity anchored to durable organization, leadership, and strategy, radical change is undercut as the summits of power are simply conceded to the ruling interests, which possess an abundance of organizational and material resources. Facing this reality, liberal and progressive groups in the United States have long fixed their sights on easy passages to social change—immediate reforms, legal rulings, technical fixes, lifestyle shifts—carry popular struggles forward but soon enough teach strategic limits. Those further to the left (anarchists, grassroots activists, Greens, etc.) are often drawn to spontaneous mass action, convinced that ordinary people need little political or ideological direction to advance deeply felt interests. "Politics," by this reading, largely serves to block the organic potential of grassroots activism. The present global crisis, however, has reached such a level of urgency and scope that confronting it without abundant political resources is destined to be futile, self-defeating. Since the planet cannot endure yet more decades of corporate predation, militarism, civic violence, and ecological devastation, conventional social routines will soon have to be questioned and overturned, vital to a transformative project strong enough to halt the global descent into "barbarism."

In the United States, hostility to politics has a long history and still infects the public sphere with special virulence, witness recurrent attacks on "big government," nightmarish references to "big brother" and "bureaucracy," worship of "free market" solutions, technological utopianism, fetishism of lifestyle changes, and the postmodern fashion of "identity politics." The seductive appeal of painless solutions and convenient shortcuts in the face of deep social problems—a familiar American escape from politics—paradoxically increases at a time when corporate-military power becomes more entrenched, authoritarian, and globalized, moving beyond the reach of democratic participation. Americans have historically been drawn to the liberal ethos of possessive individualism rooted in personal consumption, property ownership, and economic self-interest. Viewed thusly, widespread alienation from government and politics is hardly novel to the American landscape, though it has emphatically hardened in recent decades. Meanwhile, the triumph of corporate-state capitalism impedes the growth of leftist politics: indeed socialism was marginalized in the United States decades before the Soviet collapse of the early 1990s. That American Greens would adopt a more conventional *modus operandi* was therefore to be expected. Ecological radicalism, on the other hand, demands a full-scale human reorientation to nature, transformed class and power relations, and generalized challenges to domination.[25]

A key challenge of the early twenty-first century is to identify sources of the modern crisis, with the aim of showing how political arrangements and economic interests operate in tandem to block far-reaching change—that is, to subvert discourses and actions opposing business-as-usual. Elite rule is protected by dense and byzantine electoral processes, concentrated wealth, an oligopolistic economy, and hegemonic forms of ideology reproduced daily through the mass media and popular culture. This book devotes attention to the shortcomings of familiar liberal outlooks such as that elaborated by Al Gore in his well-received documentary and written work—as well as the limitations of environmental movements and perspectives that fall short of political-strategic articulation. An inquiry into radical possibilities must critically revisit the great historical legacies of Marxism, Leninism, anarchism, new social movements, Greens, even social democracy. Newer political departures will be enriched by a collective learning from successes and failures of these traditions, some of which (including Marxism) continue to exercise intellectual (if not political) influence across the globe. It is worth exploring how the Greens, for example, following in the trajectory of European social democracy, eventually succumbed to the same logic of deradicalization.

For more than a century, oppositional groups, movements, and parties have been overcome by the integrative logic of large-scale organizations, as Robert Michels famously theorized in his "iron law of oligarchy," along with the hegemonic influence of ruling ideologies—dual impediments to radical change.[26] Further, postwar transformations within global capitalism have rendered earlier anti-system models of change problematic, even where those models survive. A more globalized and concentrated state-corporate apparatus suggests tightened (though not always successful) regulation of class conflict at a time when the working class itself is more diversified as government, service, and technological sectors expand. Meanwhile, the steady growth of industrial and financial centers of power, aided by strong neoliberal policies, reduces ideals of "free markets," local autonomy, and self-governance to fictions, ideological rituals of corporate-liberal ideology.[27] Perhaps more significant, widening influence of the media, education, and popular culture raises new questions about the formation and character of mass consciousness, calling into doubt the assumption (embraced by Marxism and anarchism) that working-class self-activity innately and progressively leads to oppositional consciousness and transformative politics.

Whatever the record of its past failures, however, radical politics will have to occupy the forefront of global agendas, and soon. Political arrangements in the United States and other industrialized societies, including cumbersome electoral and legislative systems, are explicitly designed to block attacks on the status quo. To reverse the modern crisis, global-warming experts like Hansen recommend ambitious reforms as a *first step*—stiff carbon taxes, rapid phase out of coal plants, massive reforestation projects, rejuvenated infrastructure, far-reaching shift toward green energy sources.[28] Such initiatives, which ignore critical problems like food, agriculture, and water resources, are now fiercely opposed by corporate interests as a threat to growth, jobs, and prosperity. Hansen and fellow critics acknowledge that even a dramatic and imminent shift toward alternative technologies like solar and wind can at best only minimally offset accelerated climate change; more basic economic and social changes are needed. One roadblock a decade into the twenty-first century is that few governments and corporations, if any, seem ready to accept even modest emission reductions, less than what Hansen and the IPCC recommend. Research shows that global warming, if left unchecked, ameliorated by only modest reforms, or finessed by a turn toward "green" capitalism, is likely to reach cataclysmic levels in just a few decades—a specter eliciting denial and stonewalling at the summits of power. Radical change cannot take off without some combination of mass protests, demonstrations, workplace occupations, grassroots insurgencies, general strikes, and electoral campaigns, leading eventually to a rebuilding of urban centers, a shift away from suburbs, renovated farming systems, and revitalized public infrastructure fueled by sustainable energy. The public sector, in fact, devours huge amounts of energy resources, with federal, state, and municipal governments in the United States combined accounting for nearly 40 percent of domestic product, registering the biggest carbon footprint in the world. A large source of greenhouse emissions (2 percent of world total) is the U.S. military, at a time when the Pentagon remains exempt from environmental regulations and insulated from media scrutiny.[29]

Recent political trends in the United States and across the industrialized world, unfortunately, pose serious new obstacles to anti-system groups, movements, and parties. The Keynesian consensus built across several decades on liberal welfare-state programs, corporate and financial regulations, and institutionalized class partnerships is under renewed attack, giving the ruling interests more freedom to indulge their oligarchic, authoritarian, freewheeling impulses. Virulent attacks on the public sector, waged as something of a holy war by the Tea Party, most Republicans, and some Democrats, reflect protracted campaigns by the American right wing to eviscerate government funding of infrastructure, social programs, safety nets, and public regulations. Their obsession with government debt, public deficits, and tax cuts works against even timid environmental reform, largely nullified by the calculated regimen of fiscal austerity. In 2011 President Obama agreed to a "deficit reduction" plan, under intense Republican pressure, that brought unprecedented spending cuts without tax increases while setting up social security and medicare for eventual attack. Although Tea Party stalwarts were catalysts of this maneuver, their wrecking objectives converged with long-standing Republican efforts to manufacture shortfalls as a whip against social programs even as the main levers of big governments (military, law enforcement, intelligence, homeland security, etc.) remain fully intact. Never a viable, much less constructive, public policy, "deficit reduction" serves as a cynical device to eviscerate those government sectors the right wing considers ideologically objectionable.[30] Hostility to "big government" was driven by a mixture of ideological fanaticism, class warfare (from above), and defense of corporate and wealthy interests—creating new and more troublesome roadblocks for even liberal environmentalism.

If major political parties across the industrialized landscape have failed to seriously address the crisis, Republicans in the United States can boast of an especially ambitious rollback agenda, proud to be fighting even initial legislative measures to fight global warming. Beholden mostly to corporate lobbies, Republicans not only resist new environmental measures but try to subvert existing public regulations wherever possible; theirs is, first and foremost, an antiregulatory agenda serving gigantic business interests. In April 2011 every House Republican voted to overturn an Environmental Protection Agency (EPA) scientific finding that climate change threatens both the environment and public health. Republicans generally oppose even modest legislation to restrict carbon emissions. Not content with moving to sabotage the Clear Air and Clean Water Acts, they look to dismantle the EPA itself, freeing corporations to further destroy the natural habitat in the name of growth and profits. Republican consensus around environmental rollback and denial is so absolute that few congressional members have dared to challenge the lockstep ethos, tied to the familiar refrain "regulations kill jobs." Republican assault on the environment—when combined with their attack on labor, consumers, and the public sector—has few political equals in the modern world.

As in past eras, social forces backing the power structure manage to conceal their self-interested, reactionary motives behind a façade of populist rhetoric and the libertarian claims of "outsiders" railing against a corrupt Beltway "establishment"—a fiction eagerly and uncritically disseminated by the media. Strong anti-Keynesian upsurges are in fact generously funded by some of the wealthiest Americans, always ready to assist right-wingers looking to brand themselves as a grouping of ordinary folks who like to mingle at working-class bars, attend NASCAR races, and speak down-home English. Meanwhile, "blue-dog" Democrats join Republicans in stepped up efforts to overturn New Deal reforms. Here Obama's bipartisan gesture of "reaching across the isle" and taking the "middle ground," recalling Bill Clinton's presidency, should have been anticipated by close observers of the American scene. The rightward turn in American politics goes much deeper than the Tea Party insurgency or White House concessions: it is rather the outgrowth of neoliberal globalization, the debilitating role of money in politics, the atrophy of labor unions, and the expansion of corporate media culture. The prospects for genuine environmental reform thus face new impediments with each passing electoral spectacle and political-branding campaign.

Elite maneuvers to stave off the deepening crisis, meanwhile, gravitate toward consumer-driven growth, the supposed magic of market solutions, and debt-led prosperity—all bound to aggravate matters, as the assault on Keynesian public investments and protections only hastens both economic and ecosystem collapse.[31] Here short-term fiscal and monetary policies, along with federal stimulus programs, treat surface phenomena rather than more fundamental sources of crisis, whether national or global. Popular resentment toward "elites" is yet another expression of right-wing populism, recalling earlier periods of severe crisis, as in Europe between the two world wars. Instead of taking on corporate and financial power as a source of personal and social miseries, workers and the poor are encouraged to focus their rage on easier targets like foreigners, "socialists," intellectual elites, and "big government," all staples of media stereotyping. The corporate-state is thus emboldened to stave off ecological and related demands for a more active public sector, better wealth distribution, demilitarization, and stricter corporate regulations.

While Republicans and the Tea Party support untrammeled corporate and financial power behind a façade of "populism" and "libertarianism," more authentic grassroots movements (more accurately, a *chain* of movements) exploded onto the American scene in late 2011 beginning with Occupy Wall Street (OWS), spreading rapidly from New York City to dozens of American cities and even abroad. The OWS protests targeted the wealthy and privileged "one percent" on behalf of a wronged "99 percent," fixated initially on the Wall Street banks but extending to issues of disempowerment, labor rights, poverty, the environment, and militarism—the very issues that elicited little interest among righteous Tea Party insurgents. Remarkably, as of November 2011, OWS had become the site of several of the largest popular mobilizations in American history. Its "99 Percent Declaration" called for a ban on all private funding of election campaigns, elevated taxes on the rich, medicare for all, dismantling of U.S. foreign military bases, an end to wars in Iraq and Afghanistan, a freeze on home foreclosures, and *expanded* EPA powers. Like the aforementioned social movements, however, the OWS grassroots presence (understandably hostile to both major parties) remained essentially pre-political, having yet to develop any broader organizational cohesion or strategic definition.

The deep ramifications of the global crisis inevitably call into question the very structural and ideological foundations of modern capitalism—the entire matrix of international corporate power rooted in profits, growth, resource mobilization, and the ever-expanding domination of nature. As privatization and accumulation of material goods are the essence of this system, it must run headlong into the requisites of environmental sustainability, not to mention social justice and democracy. In this process economic and governmental institutions work in tandem, not in opposition, as so much right-wing thinking insists; the "national interest" is fully identified with the needs and priorities of capital. And here, as Michael Parenti notes, "Ecology's implications for capitalism are too momentous for the capitalist to contemplate."[32] There can be no ecologically sustainable capitalism, as the plutocrats view the world system as nothing but an abstraction, Parenti adding, "... they live not in the long run but in the here and now. What is at stake for them is something more proximate and more urgent than global ecology. It is global capital accumulation."[33] By the early twenty-first century it could be argued that the ecological crisis poses the greatest of all threats to the workings of globalized corporate power, as efforts to solve the crisis will demand new modes of thought and behavior. On this point Klein observes that "Climate change is a message, one that is telling us that many of our culture's most cherished ideas are no longer viable."[34] This is a system, after all, that by its very logic violates every principle of living ecosystems.

Radical politics today will have to appropriate an ecological model consistent with transforming natural relations, the system of production, social arrangements, and state governance. Could such a politics gain headway in coming decades as the global crisis deepens, everyday life is disrupted, and fresh strategic departures become real possibilities? Could such a politics be energized by the democratic and communitarian impulses of social movements that must also confront deeply entrenched power structures—a strategic imperative that consistently eluded the earlier traditions? Given the extent to which corporate power saturates the public landscape, will counter-hegemonic struggles be empowered to achieve full *political* articulation? A more daunting question: will a revolutionary breakthrough absolutely depend on local self-activity, or might some variant of Jacobinism (vanguard party) be necessary in an era when Communist parties have long been discredited? Insofar as the ecological crisis most clearly illuminates the vast global challenges ahead, how will relations between people and nature, society and environment, politics and social movements need to be reconstructed? A final, related question: what kind of political strategy is needed for a project of social governance and sustainable development on a world scale? If this book furnishes no hard and fast answers to such vexing questions, it is written with the hope that even tentative and necessarily impressionistic responses will provide enlightening points of departure in the epic struggle to avoid planetary disaster while advancing prospects for a more just, democratic, peaceful, and livable world.[35]

Chapter 2

The Global Crisis Worsens

The global crisis today is qualitatively different from historical crises, its dimensions at once economic, social, ecological, military, and perhaps above all, political. A daunting totality, the current predicament threatens continuation of life on the planet as known for thousands of years. It transcends the episodic or cyclical dynamics of earlier crises, stemming as it does from a protracted downward trajectory worldwide in scope and system challenging in gravity. Economically, the world faces worsening material and social inequalities, one-sided growth, eroded public infrastructures, decaying industrial orders, and poverty-ridden cities. Socially, there are growing instances of human dislocations, community breakdown, civic violence, and everyday turbulence as world population expands by almost 90 million people yearly and urbanization continues apace. Ecologically, the crisis revolves around resource depletion, food shortages, massive pollution, unsustainable development, and the already visible consequences of global warming. Militarily, the world is beset with mounting geopolitical rivalries, resource conflicts, and possible outbreaks of warfare. Politically, decision-making failure is accompanied by institutional breakdown and impasse, largely resulting from transnational corporate power enforced by the international organizations it controls. Despite planetary ecosystems being rapidly overrun and compromised, there are few indications that ruling elites across the globe are prepared to intervene in ways needed to reverse, or even contain, the crisis.

Earlier variants of crisis theory were usually a product of the Marxist tradition, fixated on the centrality of the economy and material relations—on contradictions between private accumulation and public needs, wage labor and capital, market and social priorities, the logic of overproduction and the fragility of consumption. For Marx and later Marxists, a mature capitalist system could never survive the explosive thrust of these contradictions, to be resolved through sharpening class struggle leading to revolution. While earlier crisis theory was often debunked or reformulated by such Marxists as Bernstein, Lenin, Luxemburg, and Gramsci, premises of the "classical" model infuse leftist thinking even today. The view of global crisis advanced here, however, is considerably more holistic in its attention to multiple and overlapping causal factors going well beyond the material dimensions of historical development. With the passing of time, references to "laws" such as the "falling rate of profit," proletarian immiserization, or capitalist breakdown have turned out to be essentially useless. Deepening contradictions within the world capitalist system can surely be identified, but their character and trajectory depart from traditional Marxist assumptions, their "resolution" (insofar as there is one) located both outside and within the unfolding class struggle between labor and capital, encompassing above all the realm of natural relations.

Point of No Return?

The world scientific consensus points toward a global ecological predicament that is steadily worsening, suggesting that the famous "tipping point" has already been reached—or will soon be reached—meaning that the crisis and its nightmarish consequences are likely irreversible. Put more bluntly, humanity appears to have reached the point of no return, where available political remedies have possibly vanished forever. Of course, for any remedy to take shape, political decision makers must recognize the urgency of the crisis—but signs of this are now remote in the United States, long the major contributor to ecological decline. While the crisis is generally framed against the backdrop of global warming, the reality is something broader: beyond climate change, it extends to such problems as resource depletion, toxic wastes, deforestation, population pressures, agricultural challenges, urban deterioration, and geopolitical conflict. The tipping point must be understood in the context of global capitalism, a system of expanding corporate power innately driven by profits and growth, resistance to public regulations, the commodification of nature, and oligarchical interests. American corporate and financial interests occupy the center of this unsustainable regimen, enforced by a massive state apparatus, war economy, media system, and worldwide deployment of military force, a matrix of domination largely outside democratic governance, unaccountable to popular demands both domestically and globally. That big business tolerates few limits to its endless accumulation of wealth and power is bringing the planet toward ecological calamity. As Senator Bernie Sanders of Vermont observed in 2009: "The very rich [in the United States] want more, more, and more and they are prepared to dismantle the existing political and social order to get it."[1] These same elites are similarly ready to destroy the ecological foundations upon which all modern economies, indeed all public life, must ultimately rest.

In the United States, Republicans—joined by many Democrats aligned with the same corporate interests—have set their sights on gutting social programs, regulations, and other barriers to unfettered economic growth. The legitimating ideology is a renovated but phony "libertarianism" embraced by corporate chieftains, the Tea Party, and a media culture ostensibly dedicated to freedom and individualism but in reality firmly attached to the centers of power: megacorporations, Wall Street, the Pentagon, national security-state. Reflecting on this deceit, Sanders adds, "The billionaires and their supporters in Congress are hell-bent on taking us back to the 1970s, eliminating all traces of social legislation designed to protect working families, the elderly, children, and the disabled. No 'social contract' for them. They want it all."[2] The ruling interests seem perfectly ready to dismantle such bodies as the Environmental Protection Agency and the Occupational Health and Safety Agency, not to mention legislation like the Clean Air and Clean Water acts passed decades ago to curb the most harmful effects of modern industry. Toward this end huge business interests and their lobbies spend billions of dollars on public relations, advertising, and interest group peddling while also reaping the benefits of corporate media. Meanwhile, as global natural habitats continue to deteriorate, with no end in sight, such behemoths as ExxonMobil and Shell amass record profits in the tens of billions while paying executives with incomes and super-bonuses worth tens of billions more.[3]

The power elite now enjoys greater political leverage than at any point in recent American history, its flexibility bolstered by a perpetual weakness of countervailing or anti-system forces. The global terrain for capital relative to labor, local communities, social movements, and weaker national governments has been largely cleared of nettlesome obstacles. If routine business for corporations and governments is now suicidal for the planetary ecology, elites remain hell-bent on a profit-making agenda that trumps all other priorities, at a time when, as Bill McKibben notes, "Global warming turns the idea of 'development' into a cruel joke."[4] The resulting plight, McKibben adds, means that "We don't really know were to turn because the planet we now inhabit doesn't work the way the old one did."[5] Efforts to simply "reform" the insatiable growth machine, worshipped by liberals and social democrats as well as conservatives, are laudable but will do nothing to stem the downward trajectory.

World ecosystems are in such deep peril that familiar political remedies may already be outmoded—or, put differently, will have to be superseded by more ambitious, far-reaching, indeed, radical solutions. A new political vocabulary is needed but has been elusive. In the United States, the ruling interests have long been preoccupied with discourses of bigness, maximum growth, the "market," private consumption, and mastery of nature—all supposedly crucial to the nebulous American Dream. By the early twenty-first century, this model, having brought continuous (but highly uneven) growth and prosperity, had turned into a darker version of the Enlightenment, a source of civic decay, alienated politics, and ecological destruction. Growth and prosperity turn ever more dystopian, fueled by a system of production and consumption fully at war with nature, robbing from the earth's natural systems at rates never before witnessed or scarcely imagined. Since World War II American society has undergone a dizzying pace of change marked by dynamic industrial growth, public infrastructural expansion, nonstop suburbanization, debt-driven consumerism, recurrent scientific and technological breakthroughs, an increasingly mobile and educated population, and a sprawling media-entertainment complex. This system was (and is) powerfully driven by globalization—economic, political, military—undergirding the largest and most potent empire in history. By the late twentieth century the American economy was devouring fully 25 percent of world energy resources while contributing roughly the same levels of waste, pollution, and greenhouse emissions from only 5 percent of the earth's population. This was a society, however, destined to eventually find itself in a downward spiral of deep financial crisis, industrial decline, shrinking public infrastructure, social atomization, and militarism, all the while defiantly and tenaciously holding to the same unsustainable course.

In his rather apocalyptic book, *The Long Emergency*, James Howard Kunstler argues that Americans are sleepwalking into the future, facing imminent disaster with a mixture of cavalier indifference and collective denial—attitudes obviously at odds with prospects for serious remedies. The global crisis has so far been met with little political awareness of such issues as food shortages, end of cheap oil, spread of new (and old) diseases, climate change, an increasingly toxic planet, dwindling energy supplies, the widening gulf between rich and poor, and mounting social chaos and political instability across the globe.[6] Population pressures can be expected to further burden the earth's carrying capacity, aggravating all other problems. As the world system continues along its existing path, with no contraction in growth, the crisis can only worsen, approaching the conjuncture when political solutions, however radical, simply cannot work. Beyond the mortal threat of global warming, the crisis is aggravated by multiple and interconnected environmental challenges with growing food shortages at the top of the list. Peak oil augurs an especially intractable problem given the paucity of alternative energy sources on a world scale. As Kunstler notes, "The world oil production peak [likely reached by 2010] represents an unprecedented economic crisis that will wreak havoc on national economies, topple governments, alter national boundaries, provoke military strife, and challenge the continuation of civilized life."[7] An apocalyptic scenario? Perhaps, but with the specter of resource wars (like those already visible in the Middle East) and energy shortages imminent, the question could be not so much "if" but "when" Kunstler's dystopic future is likely to materialize.

With the era of cheap and abundant fossil fuels coming to an end, Americans have entered what Kunstler calls a "consensus trance," substituting for political action an array of technological fixes, "market" formulas, and other bromides compatible with continued business-as-usual. For corporations, banks, mainstream politicians, and established media the familiar mantra remains, "bad news is not good for business," suggesting that elites have little incentive to acknowledge, much less engage, pressing ecological realities. American citizens have for decades been addicted not only to oil but to the everlasting myth of corporate goodness and efficiency—an outlook fostered daily by the media culture. As the planet encounters shrinking oil reserves, with prices surging, the American thirst for fossil fuels continues apace, as if such natural resources were infinite: food production, suburban lifestyles, the auto culture, shopping malls, the petrochemical industry, and of course the Pentagon all remain heavily dependent on carbon energy. Industrializing nations like India, China, Indonesia, Brazil, and Mexico now compete for global oil reserves, which could be smaller than officially claimed by such producers as Russia, Mexico, Saudi Arabia, Kuwait, and Iraq. The gap between new discoveries and escalating global demand is widening rapidly, with existing reserves more difficult and costly to retrieve. As is well known, the United States currently (2011) imports roughly 70 percent of its voracious energy needs—an amount bound to increase dramatically, further exacerbating global resource competition. Meanwhile, U.S. leaders continue to sleepwalk, pretending that the epoch of cheap oil and all that it nurtures will go on indefinitely, while future energy alternatives are celebrated but receive stingy amounts of funding. At the 2010 worldwide level of oil consumption of 30 billion barrels, global reserves cannot support such demand for more than another decade.

The threat to planetary stability and livability posed by dramatic weather pattern changes—in 2011, for example, no less than 19 countries set all-time high temperature records—has unfortunately been met with precious little intellectual or political sense of urgency. Polluters and their ideologues adhere to a business-as-usual regimen as the public sector comes under increasing siege from right-wing interests that defend polluters while rejecting climate change evidence as so much "liberal theory." Democratic leverage in the United States erodes as corporations and their interest groups manage to colonize the public sphere, with indispensable help from the media. Neither the political nor the media establishment devotes much attention to the great dangers of global warming, ever fearful that their growth agendas might be upended. The need for drastic (and quick) reductions in worldwide carbon emissions is rarely addressed or, if so, just casually: the Congress, the White House, political candidates, journalists, medical elites, and religious groups maintain a steadfast posture of silence. No significant environmental legislation has been passed during recent presidencies, including Obama's. A major Republican candidate for the White House in 2011, Texas governor Rick Perry, wrote a book entitled *Fed Up!* in which he attacks climate change science as a "phony mess"—a motif he repeats tirelessly on the campaign trail. With Texas hit by months of severe heat, drought, and wildfires, destroying crops and livestock at breathtaking levels, Perry urged citizens to simply "pray for rain." In nascent global-warming lore, there are the total deniers, comprising most Republicans; the scientific "skeptics," who fault research credibility and mathematical calculations; the "lukewarmers," who recognize heating trends but dismiss their significance or reject the idea of human involvement; the "sunspotters," who agree that the planet is getting hotter but blame natural phenomena (solar flares, etc.) rather than humans; and the religious deniers, who argue that climate change was never mentioned in the Bible—or, alternatively, that such change portends the "end times." Whatever the source, the ultimate price of denying the crisis will only be a tragic hastening of its dreaded consequences.[8]

The gravity of climate change and related problems is thus scarcely acknowledged at the summits of American power. There are laughable big-business moves toward greenwashing (forging an *image* of change, marketing of "green" commodities, etc.) and reforms that do nothing to disturb neoliberal routines. If, as Lester Brown concludes, the earth's carrying capacity was already exceeded in the early 1980s, global warming has simply accelerated that trend over the past few decades.[9] Elevated threats of climate change, recognized in credible scientific and governmental reports, elicit at best perfunctory nods of concern among enlightened leaders while generating the most tepid legislation. The world has experienced 25 of the warmest years on record since 1980, interpreted by respectable scientific opinion as resulting from heat-trapped greenhouse gases produced by fossil fuels. Atmospheric concentrations of carbon dioxide rose from 280 ppm in 1960 to an alarming 390 ppm in 2008, with more drastic elevations on the horizon.[10] In 2008 alone some eight billion tons of carbon were poured into the earth's atmosphere. As these patterns continue, within several decades the planet's overall temperature will rise several degrees Fahrenheit, a trend destined to trigger global calamity measured by severely diminished crop yields, melting glaciers, rising ocean levels, extreme droughts and floods, worsening storms, more frequent wildfires, and vanishing species. Studies reveal that if the current trajectory persists, with carbon dioxide levels reaching 450 ppm and higher, we face potential global Dust Bowl conditions like those of the 1930s Midwest—in other words, a mostly unlivable planet.[11] The catastrophe in this case would be "holistic," an outgrowth of combined economic, social, political, and military as well as ecological conditions. Brown concludes, "Climate change poses a threat to our civilization that has no precedent. A business-as-usual energy policy is no longer an option."[12]

The crisis is sure to be most deeply and immediately felt in the sprawling, congested, conflict-ridden megacities of the world—urban centers like Mexico City, Jakarta, Bangkok, Sao Paolo, Cairo, Shanghai, Moscow, Mumbai—all heavily dependent on complex, resource-demanding public infrastructures, where the lives of hundreds of millions of people hang in the balance. As of 2010 roughly 3.5 billion people resided in these cities, already overpopulated and burdened far beyond sustainability. There are presently 430 cities with at least one million people and 19 megacities with more than ten million, and of course these metropolises continue to expand. Mexico City, Shanghai, and Tokyo are home to more than 25 million people each, large numbers crowded into super-congested and vulnerable centers. Cities are riddled with poverty, unemployment, homelessness, and crime, while most have inadequate public services (including health care and education) for all but the most affluent. Governments face surging problems of food, water, transportation, housing, education, and public health in settings where both natural and material resources are commonly strained beyond capacity. There are rising levels of violence, disease, malnutrition, and social atomization—festering problems likely to intensify as ecological pressures mount. Population growth sharpens the crisis as it places further burdensome demands on resources and infrastructure—a point more fully developed in the postscript.

The world has already been swept into rapidly worsening weather-related calamities, one feeding on another, nowhere more than in the United States, which in 2011 endured unprecedented drought, storms, flooding, and wildfires across wide expanses of the country. Record high temperatures were recorded in New York City, Boston, Philadelphia, Baltimore, and other urban centers along with such states as Texas, which sweltered under continuous 100 degree Fahrenheit temperatures and faced hundreds of fires during the summer of 2011. The National Climate Data Center reported that Texas in summer 2011 had endured the highest average three-month temperature (almost 87 degrees) recorded in any state since government records were kept, dating back to 1895. Such weather raised fears of a return to Dust Bowl conditions of the 1930s. More than a thousand tornadoes struck areas of the south and Midwest, killing hundreds, destroying vast areas of land while leaving one entire city (Joplin, MO) virtually leveled. Other storms left wide belts of devastation, from south to northeast, floods inundating dozens of towns and cities as well as farmland. Wildfires from extreme heat and drought increased in the west and southwest at a nearly 60 percent rate. Could such phenomena be entirely accidental or the result of natural cycles, as many warming deniers claim? In fact the IPCC by the 1990s predicted such extreme weather events as an outgrowth of global warming—of heat-generating carbon emissions and rising sea-surface temperatures measured in the Atlantic and the Gulf of Mexico.

Events in the United States coincided with those around the world, as heat waves swept Europe, Russia, and many parts of Africa, leaving in their wake drought and parched landscapes. Rainfall across Africa dropped by up to 90 percent, inflicting on such nations as Kenya, Ethiopia, the Sudan, and Somalia what UN monitors called "the worst humanitarian disaster" of the period, leading to declining agricultural output, famine, severe water shortages, and massive flow of refugees in search of food and water. Could this prefigure more such disasters ahead? Cities of the world now stand as cauldrons of risk from potential catastrophes, both natural and human-made. Not only are huge urban centers vulnerable to episodes like earthquakes, tsunamis, hurricanes, and floods, their public infrastructures are scarcely prepared to handle the daunting material, environmental, social, and public health challenges that surely lie ahead. Many cities—rapidly growing, congested, and overburdened—lack adequate public facilities and policies to face the immense challenges of climate change, not to mention the disasters stated, while the corporate and financial centers of power (largely transnational) are often corrupt and generally unaccountable to the public. Meanwhile, as the global economic crisis imposes new fiscal restraints on national and municipal governments, infrastructure funding will continue to lag even further behind public needs.

As scientist James Hansen argues in *Storms of Our Grandchildren,* human capacity to plan for the future is overwhelmed by the primacy of short-term corporate interests, especially in the United States. The prospect of global calamity is scornfully denied by those in power and by the affluent, where life seems to proceed more or less on a normal course, absent any sense of urgency. With empirical indicators of the crisis hard to overlook, the gulf between ecological realities and public awareness of those realities is actually widening. One case in point is the threat of species extinction, accelerating at the rate of several per day, its future implications more harrowing than commonly believed. Hansen writes, "I will argue that if we continue a business-as-usual path with a global warming of several degrees Celsius, then we will drive a large fraction of species, conceivably all species, to extinction."[13] Yet, despite graphic images of severely challenged life forms everywhere, this issue scarcely enters American public discourse at any level. This is hardly surprising, as global-warming denial is pervasive throughout American political and intellectual life. One source of denial is that oil, natural gas, and coal still furnish about 90 percent of American energy sources. Energy corporations are little inclined to shift away from what do date have been relatively cheap, profitable fossil fuels, or to favor policies (like green transportation) that lessen reliance on carbon-emitting energy. Thousands of well-paid, aggressive lobbyists in Washington, D.C., and state capitols work tirelessly to protect giant energy interests, scarcely concerned about the grave ecological consequences. Hundreds of scientific papers, environmental reports, and groundbreaking books have sadly exerted negligible political impact. Hansen's own dire warnings were scandalously undercut by the stonewalling Bush administration (and oil presidency), with Hansen himself eventually forced to leave.

Political escapism, however, will do nothing to stave off environmental horrors and in fact promises to make them worse. Hansen warns, "Business-as-usual greenhouse gas emissions, without any doubt, will commit the planet to global warming of a magnitude that will lead eventually to an ice-free planet. An ice-free planet means a sea-level rise of about 75 meters (almost 250 feet)."[14] Of course, long before that catastrophic moment the world is likely to sink into a deepening Hobbesian phase of chaos, breakdown, and anarchy, a future too dystopic to contemplate outside the realm of Hollywood horror films.

Against this backdrop, the well-circulated 2007 report of the IPCC (based on reports assembled by the United Nations) actually *understates* the scope of the crisis, predicting, for example, relatively modest world temperature and sea-level rises by the end of the twenty-first century.[15] The Panel did receive a mountain of criticisms attacking its relative conservatism and alleged pandering to corporate interests, even after being awarded the Nobel Peace Prize. In fact the bulk of data gathered by the IPCC turns out to be rather dated in its accuracy and projections. Crucially, the Panel refused to place blame on gigantic oil and coal industries that, by all calculations, are responsible for the bulk of greenhouse gases. Still, a signal IPCC contribution was to legitimate and solidify the worldwide scientific consensus regarding *human* contributions to global warming. A team of UC Berkeley scientists, believing that they might find gaping holes in the scientific consensus, conducted research based on data from hundreds of weather stations around the globe—but their conclusions, reported in 2011, actually *confirmed* those trends identified by the IPCC. The Berkeley Earth Surface Temperature project, funded in part by billionaire and warming denier Charles Koch, aimed to correct a presumed "bias" of the international scientific community but wound up lending more credence to that "bias."

The annual total of carbon emissions attributable to human activity (in 2010) was roughly 37 billion tons, about 25 percent of it generated in the United States—an amount figuring to rise steadily along with population growth and the expansion of cities, factories, power plants, office buildings, residential areas, and food production not to mention autos and their infrastructure. In 20 years this number could reach 60 billion tons or more. Aside from carbon dioxide, other pollutants such as methane, nitrous oxide, and aerosols enter into the climate change equation, but carbon is the primary culprit as it remains in the atmosphere for decades, even centuries. The carbon level of roughly 390 ppm has reportedly not been reached for millions of years. On a per capita basis, emissions are presently

25 tons yearly for the United States, 11 tons for Europe, eight tons for China, three tons for India, and one ton for the African continent. Mounting scientific data suggests that carbon stabilization (essential to sustainable development) would require the world, on average, to shrink to the poorest (African) level, and very soon. That would mean a staggering 95 percent reduction in overall American consumption. The ecological (not to mention the ethical and political) challenge posed by such an imperative requires no further elaboration. A critical problem for the United States is that the bulk of fossil-fuel utilization stems less from industrial production than from a variety of lifestyle patterns—the auto culture, suburban living arrangements, air-conditioned offices, meat-centered agriculture, the fast-food industry, a sprawling media-entertainment complex. Here it is worth noting that the U.S. military alone is the biggest single cause of greenhouse emissions in the world—hardly surprising, given its hundreds of bases scattered around the world, its fleet of tens of thousands of vehicles, and the warfare it has been conducting in the Middle East. The Pentagon consumes nearly 350,000 barrels of oil daily. Its operations destroy land, agriculture, water sources, and public infrastructure. It is the source of massive toxic wastes in the form of carbon pollution, toxic chemicals, residual land mines and cluster bombs, and radiation from depleted uranium as well as nuclear deployments. Unfortunately, the U.S. military has long remained unaccountable, exempt from climate change agreements along with domestic environmental constraints on its operations.

The global predicament takes on a new meaning as nations like China, India, and Brazil seek a level of modernity associated with the affluent (and unsustainable) lifestyles of advanced capitalism. Meanwhile, carbon levels rise by roughly two ppm yearly, which indicates that the aforementioned 450 ppm will be reached in three decades if not sooner.[16] No one knows precisely when the fearsome tipping point will arrive, but with each rise in carbon emissions imminent global catastrophe becomes more likely.

Previous estimates that carbon levels must be contained at 450 ppm to avoid global collapse now appear rather fanciful. Hansen for one argues that we have already reached the stage where ecological disaster will be difficult to avoid, that a level of 450 ppm is indeed untenable, and that a more reasonable level is closer to 350 ppm—far below present concentrations.[17] Hansen writes, "While some of the public is just becoming aware of the existence of global warming, the relevant scientists—those who know what they are talking about—realize that the climate system is on the verge of tipping points. If the world does not make a dramatic shift in energy policies over the next few years, we may well pass the point of no return."[18] This shift, however, will require far more than altered energy policies, involving transformed modes of production and consumption. Moreover, emissions from current and earlier human activity will linger in the atmosphere for many decades once they are reduced—that is, global warming from existing sources might take several decades to peak. If the crisis veers out of control, with most or all planetary ice gone, sea levels inundating vast areas of land, and food production drastically compromised, the momentum could be much too strong for even the most far-reaching measures to reverse. In that case, as Hansen predicts, the planet will move onto the fast track toward the Venus Express, eventually rendering it all but uninhabitable.[19] And, with continued business- and politics-as-usual, that future is probably not too distant.

The unfortunate reality is that humans are rapidly losing the political capacity to manage—much less reverse—this course of events; impasse leads inexorably to impending catastrophe. In other words, there is the question as to when effective solutions to the global crisis will no longer be possible. It could be that humanity—or, more concretely, the predatory interests behind capitalist modernity—has laid the path of its own destruction. The Enlightenment tradition, which promised so much in the way of human empowerment, appears to have turned on itself, a casualty of the great myth of progress through heroic advances in science, technology, and industrial growth. The epic ideals of freedom and democracy, not to mention prosperity, dependent on such "progress" could become a dream turned nightmare, the dark legacy of corporate power, technological rationality, bureaucracy, and militarism. Here McKibben aptly notes that "Global warming turns the idea of 'development' into a cruel joke."[20] This manifestly dystopic side of the Enlightenment is what the power elite cannot face, and what the general public has been propagandized to avoid. All signs point to the U.S. power structure desperately holding to its outmoded system of production, driven by fossil-fuel addiction, endless growth, privatized consumption, and globalized reach. This model remains seductive for, as Kunstler observes, "Oil led the human race to a threshold of nearly godlike power to transform the world."[21] Such "godlike power," we have come to realize, has begun to devour itself, threatening to demolish its own small grouping of architects along with its many victims. In American society, perhaps more than elsewhere, a one-dimensional growth machine continues to replicate the worst features of modernity: deregulated corporations, larcenous banks, a war economy, meat-based agriculture, suburbia dependent on

The Coming Food Nightmare

Nothing is more likely to dramatize (and exacerbate) the global ecological crisis than the decline of world agriculture and, with it, increasing shortages in food production and distribution. Here a confluence of factors—resource shrinkage, pollution, developmental limits, social traditions, and climate change—is destined to surface. Warning signs are already ominous, as the world enters a food era in which most nations are running out of arable land sufficient to feed their populations. By 2010 more than one billion people faced severe food problems and malnutrition, with perhaps another billion in line to suffer the same fate while affluent countries of Europe and North America were still living in dietary splendor if not excess.

The American public was told until recently that the modern food system—industrialized, commercialized, high-tech, chemicalized—was a great triumph of human ingenuity, a signpost of capitalist achievement. For advanced industrial nations, the endless supply of meat, dairy products, fast foods, and agricultural goods has long been a birthright, the product of a smoothly efficient machine needing far less labor power than traditional agricultural systems. These goods were readily available, cheap, and, for the most part, highly satisfying for ordinary consumers. Such plagues as malnutrition, not to mention starvation, had apparently become fears of the past. Beneath these comforting sentiments, however, a global agricultural crisis was unfolding and could not be denied or rationalized away, even in the centers of food abundance. As of 2011, the world suffered declining food production capacity relative to demands of a growing population that could reach 7.5 billion or more (from the existing 6.5 billion) by 2030. With planetary carrying capacity likely exceeded a few decades ago, the food issue now merits priority on the global agenda. While the number of inhabitants increases, the world's arable land is mostly already in use, and much of that land (owing to urban encroachments, massive soil erosion, over-chemicalization of crops, water shortages, and drought) will shrink in coming years. Under such conditions, perpetual food abundance—assuming present developmental patterns—will soon turn into a sad mirage. In his prophetic book, *The End of Food*, Paul Roberts writes, that "... tomorrow's high-tech farmers will face a very different world. Not only will they be trying to feed more people but they will be doing so without the benefit of three critical advantages their predecessors took for granted—cheap energy, abundant water, and a stable climate."[22] While insightful, Roberts seems to have overlooked perhaps the most decisive factor: productive arable land.

The best evidence suggests that world food output will fall precipitously in coming decades, precisely when aggregate demand is sure to grow dramatically. One problem is that agriculture everywhere has become more vulnerable to drastically shifting weather patterns: drought, heavy rains, extreme heat, and flooding, as mentioned earlier, have already begun to wreak havoc on crops and livestock support systems. Climate change augurs eroding water supplies, already in jeopardy with elevated demand and the draining of underground aquifers across the planet. Declining oil supplies—leading to skyrocketing energy costs—will undermine food production, from harvesting to processing, distribution, packaging, and sales. Arable land, as noted, is declining because of soil erosion and toxic contamination, dry and hot weather, and both residential and commercial expansion. In coming years large regions of the planet (including sections of China, Africa, and Latin America) will be hard put to harvest adequate crops or support livestock at levels needed to feed larger populations. The world agricultural system is already unsustainable, emphatically so when resource-depleting, costly, meat-centered diets are taken into account. Animal products even now consume more than one-third of all grains harvested worldwide. The food crisis is compounded by an agribusiness economy dominated by multinational corporations such as Cargill, Nestle, Kraft, Tyson, Monsanto, Wal-Mart, and McDonalds, all geared to industrial processes that are costly, unsustainable, and resistant to change.

As elsewhere, the United States occupies a special niche within this developing food nightmare, especially since meat and fast food have become central to the American economy and culture, now among the great national addictions. Americans presently consume a staggering 217 pounds of meat per capita yearly, compared with just 12 pounds for India and even less for most developing countries. Livestock support systems in the United States devour nearly 70 percent of all grains, 50 percent of all water, and 10 percent of all fossil fuels (used in fertilizers, pesticides, and food processing). In this context, an American-style diet would be sustainable for barely two billion people, less than one-third of the *existing* world population. To meet future food sustainability, the United States would have to reduce its meat intake to something like the Indian level (roughly 5 percent of current levels), but the United States and most would-be affluent nations are moving in the opposite direction. By 2030, total world meat demand is projected to rise by 70 percent, exerting unfathomable pressures on water, petroleum, soil, and land resources.[23] As nations of the world seem hell-bent on emulating the wasteful and destructive American food model, even mainstream observers like Roberts are forced to argue, "And the conclusion is fairly stark: under any model for a future food system that is both sustainable and equitable, the meat-rich diets of the West, and especially of the United States, simply don't work on a global scale."[24]

As noted earlier, the planet has reached—or will soon reach—the threshold of peak oil, meaning that energy prices will steadily rise as other resources grow more scarce and expensive. Industrialized, meat-based agriculture requires a constant flow of cheap and abundant fossil fuels. Oil prices increasing from $20 to $95 and more per barrel had by 2011 already created havoc with food production, distribution, and processing in the United States, Europe, and elsewhere. Of course, peak oil means rapidly increasing costs of doing business in every sector. World oil output is projected to drop from 85 million barrels daily to 55 million barrels by 2020, signaling a price elevation to $150 a barrel or more—a disaster scenario for global agriculture as presently organized. What will be the economic and political consequences when oil reaches $150 or even $200 a barrel? Peak oil ensures that world populations will confront a food crisis of cataclysmic proportions, exacerbated by converging pressures like water shortages, soil erosion, declining arable land, and climate change.

As with other spheres of "personal life" now generally understood as having wider public relevance, meat has long been regarded as a "private" issue, a matter of individual dietary choice belonging outside the scope of social policy; the issue of animal-based farming and fast-food production has therefore been largely off the political radar. In the United States, past decades have witnessed *some* changes in popular attitudes toward meat, yet most Americans see no connection between meat consumption and a broad range of social and individual problems. These problems, as we have seen, are indeed plentiful: resource depletion, air, water, and soil pollution, food shortages, deforestations, disease, global warming. *Worldwatch* magazine has observed, "... as environmental science has advanced, it has become apparent that the human appetite for flesh is a driving force behind virtually every category of environmental damage, including the growing scarcity of fresh water, loss of biodiversity, spread of toxic wastes and diseases, even the destabilization of countries."[25] This plight is aggravated by a secular increase in global demand for meat in recent decades: with 6.5 billion people on the planet in 2011, at least 90 percent consume meat products regularly while the absolute number of meat consumers increases yearly. A source of astronomical profits for agribusiness, meatpackers, grocers, and the fast-food industry—indeed a major bulwark of the entire corporate system—meat today in its myriad forms is a decisive factor in ecological deterioration, not to mention chronic health afflictions.[26]

Animal-food production in the United States alone has risen fourfold since the 1950s, despite the spread of scientific knowledge revealing the harmful effects of meat eating, including cancer, heart disease, diabetes, osteoporosis, and other chronic ailments. At present there are an estimated 22 billion livestock on earth. In the United States more than 100,000 cows and calves are slaughtered every day, along with several million chickens. The Tyson food complex at Noel, Missouri, kills some 300,000 chickens daily while the IBP slaughterhouse at Garden City, Kansas, and the ConAgra complex at Greeley, Colorado, both kill more than 6,400 steers a day.[27] All told, some 25 million animals are slaughtered each day worldwide to satisfy escalating human demands for flesh. In a McDonaldized society Americans eat on average 30 pounds of beef yearly, with seemingly little concern for well-known health risks. Conditions of factory farming, said to be improved in the wake of reforms, have in fact gotten worse by most reasonable standards—more crowded, more painful to animals, more disease ridden, more drug saturated, and more pollution causing even than at the time of Upton Sinclair's *The Jungle* (written in 1906).[28] More than half of all pigs, chickens, ducks, and other animals are afflicted with diseases like cancer and leucosis at the time of slaughter. Meanwhile, the Federal Humane Slaughter Act has done little to diminish the assembly-line terror of slaughterhouses.[29] In fact the meat industry has virtual *carte blanche* to do as it pleases with its commodities since government monitoring ranges from sporadic to nonexistent—a situation that, as John Robbins argues, amounts to a crime not only against animals but against nature.[30] That such practices are so routine, so concealed from public sight, so distant, and so ideologically sanitized hardly minimizes the horrors, or the damage. Michael Pollan, in *The Omnivore's Dilemma,* makes a telling point: "The industrial animal factory offers a nightmarish glimpse of what capitalism is capable of in the absence of any moral or regulatory constraint whatsoever."[31]

As part of the general, relentless assault on nonhuman nature, the meat-industrial complex is tied to the same corporate-imperial logic that fuels the ecological crisis, including militarism, resource wars, global poverty, and workplace authoritarianism. That animal-based agriculture is so grossly unsustainable makes little difference to those profiting from it, as all the waste and destruction is accompanied by huge corporate gains. It is of little consequence to those in power, therefore, that cattle, sheep, and other livestock graze more than 525 million acres of land, or nearly two acres for every person in the United States.[32] In all, the percentage of grains fed to livestock in the United States is a massive 70 percent (80 percent in the case of corn).[33] Further, great expanses of land, both in the United States and worldwide, have long been overgrazed, leading to soil erosion as vast regions are being deforested to make room for extended animal grazing and farming. As for global warming, livestock account for more than 20 percent of world methane emissions, not including fossil fuels used in agriculture, transport, and processing. While the enormous material waste and ecological harm brought by the meat industry is no secret, the reality is that as societies develop economically and the middle strata expands, meat consumption tends to increase sharply as it is widely considered a symbol of affluence, progress, and good living. Unfortunately, the public demand for meat and dairy products—stimulated by ambitious marketing and advertising—increases at the very moment that arable land is shrinking, oil resources are peaking, soil is being depleted, and water supply is becoming more problematic. Yet spreading awareness that animal products are twelve times less efficient in generating food for hungry populations resonates little across the political system.

Spurred by unfettered corporate expansion, neoliberal globalization—enforced by the World Bank, the IMF, and the WTO—subverts the ecological balance by its very logic, tied as it is to profits and growth. The present global order legitimates unsustainable practices on a foundation of "free markets," growth-based prosperity, and technological efficiency, claims possessing less validity as we move into the age of a worldwide food crisis. At the core of neoliberalism is an instrumental approach to nature that, over time, becomes increasingly self-destructive, untenable. If sustainability requires ecological balance—respect for nature, limits to growth, shift toward renewable resources, stewardship of the global commons—more integrated relations between humans and the natural environment are essential. Despite the prevailing wisdom that population growth matters little, it is hard to ignore potential consequences for the natural habitat of a world population that could *double* over the next 50 years. Population growth of nearly 90 million yearly means a steady decline in per capita resources, elevated pollution levels, more urban congestion, overburdened public infrastructures, new biodiversity challenges, and continued deforestation. Food and water resources will be further taxed. David and Marcia Pimentel argue that "Humanity is approaching a crisis point with respect to the interlocking issues of population, natural resources, and sustainability."[34] If planetary carrying capacity can reasonably be established at roughly 2.5 billion people, as the Pimentels conclude, it seems evident that a population of 7.5 billion will only speed up arrival of the global tipping point.[35] And if meat consumption increases at a rate close to present levels, as anticipated, the crisis will be further aggravated. In this scenario grain harvests will have diminished greatly relative to demand, the result not only of resource shortages and accelerated climate change but also of cropland loss to expanding urban centers.[36]

American society has grown disastrously addicted to meat and all that surrounds it, including the largest fast-food culture in the world. Habituation occurs and is reinforced on several fronts—political, economic, personal, even religious—reproduced by powerful agricultural, manufacturing, and service interests: fast-food output has skyrocketed, reshaping the entire American landscape.[37] According to Eric Schlosser, Americans spent $134 billion on fast food alone in 2000, more than was spent on college education, personal computers, and new cars.[38] By 2010 such expenditures had reached nearly $200 billion. Today animal products fuel the modern industrial system, having become a (false) symbol of prosperity but also a major cause of social, workplace, health, and environmental ills.

Jeremy Rifkin shows how the beef industry, heralded as a symbol of modernity and progress, evolved historically alongside an Enlightenment project dedicated to the systematic commodification of nature.[39] There is nothing in the Marxist tradition or its legacies that seriously addresses this phenomenon. Scientific discovery, technological innovation, and industrial growth were all harnessed to sprawling meat enterprises that in the United States were especially valorized within the frontier ethos. During the Westward push meat emerged as a dominant economic and cultural force, reinvigorating the capitalist ethic of rugged individualism.[40] Over time the cattle system, celebrated in hundreds of Western books and movies, became a pervasive feature of the social order, a staple of the American diet, a source of bountiful profits, and a cause of horror for animals that Sinclair was the first to bring to American public attention. By the 1950s meat figured strongly in the rise of suburbia, the auto culture, and a media system that fueled McDonaldization, a food regimen suited to fast-paced, mobile urban and suburban lifestyles as the outlets took on distinctly Fordist operating principles: uniformity, speed, efficiency, standardization, affordability.[41] All components of animal farming and meat processing were thoroughly rationalized, simultaneously generating and satisfying public demand for hamburgers, hot dogs, steaks, luncheon meats, and related fare. Workers at factory farms, slaughterhouses, canning plants, and fast-food outlets were recruited from mostly low-wage minority labor and subjected to alienating, routinized, toxic, and dangerous jobs. As for cattle, they were (and are) dehorned, castrated, injected with hormones and antibiotics, sprayed with insecticides, and transported to killing plants, then broken down into countless marketable parts, ultimately to wind up at butcher shops, stores, and restaurants. Used in literally hundreds of industrial and food products, beef alone generates huge profits for corporations like ConAgra, Cargill, Tyson, IBP, and McDonalds. The same ritual is repeated for chickens, ducks, pigs, sheep, turkeys, and other animals, by the millions each day, all subject to similar assembly-line horrors. Public awareness of the destructive consequences of meat for human health, the environment, and animals, though widening, has scarcely made a dent in this apparatus.

As McDonaldization appears to symbolize modernity in the food economy, meat becomes an eminently saleable commodity for firms benefiting from mobile lifestyles dependent on fast-paced routines and cheap energy sources. Champions of advertising and marketing, the meat companies resist government regulation in a sector most desperately in need of it to monitor health threats, toxic emissions, harsh working conditions, and extreme cruelty to animals. The industry is a bastion of right-wing politics fueled by neo-Darwinian ideology, union busting, hostility to wage increases, draconian work relations, and fierce opposition to public regulation.[42] Schlosser's vivid account of life at ConAgra's giant plant near Greeley, Colorado, reverberates with terrible narratives out of *The Jungle*. There hundreds of thousands of cattle are squeezed together in huge feedlots, so close that they can barely move, handled as dispensable units of production, and sent to their death. Animal wastes, toxic runoffs, and chemical emissions fill the slaughterhouse, spreading disease to cattle and humans alike, while workers are powerless cogs in a rationalized machine. This uniquely "American" contribution to food production -and social relations—is being exported to every corner of the globe. McDonalds has emerged as the leading symbol of a globalized meat empire, the embodiment of unsustainability now thoroughly integrated into the prevailing American lifestyle.[43]

Aside from the military, no section of American society matches the frightening consequences of the meat complex: ecological ruin, food deterioration, routinized violence, workplace injuries, disease, and death to both humans and animals—all part of virtually unaccountable corporate power. The health of consumers habituated to foods loaded with fats, salt, sugar, and calories worsens continuously, reflected in the growing incidence of obesity connected to such afflictions as cancer and heart disease. As noted, public knowledge of health problems stemming from a meat-based diet has recently increased, thanks to a new generation of critics and such documentaries as *Diet for a New America, McLibel, Supersize Me!, The Corporation,* and *Food, Inc.* No doubt too the alarming scope of American health ailments, even among youth, has given rise to something of a backlash, as reflected in the sharp increase of vegetarianism. In response, the meat industry has stepped up its propaganda trumpeting the great benefits of beef and poultry. Consumers are told, falsely, that meat is essential to good health, that it is an indispensable source of protein and other nutrients, that vegetarianism is a silly and harmful fad, that "barnyard" animals are treated with great care, that critics of meat addiction behave like "food dictators" and "lifestyle Nazis." People are warned against the sinister and intrusive schemes of a fascistic "culinary police," with fears of big brother taking over the kitchen. Lobbies such as the National Cattlemen's Association, the American Meat Institute, and the National Dairy Council, joined by such corporate-friendly "diet" regimens like that propagated by Robert Atkins, wage multibillion dollar media counteroffensives to persuade Americans that meat is the (only) path to true health.[44]

The animal-based food system is so deeply entrenched in American history and culture that serious departures from the past now seem unimaginable. Challenges to meat eating, moreover, can be taken as an insult to personal rights and freedoms associated with sensitive religious, national, and ethnic traditions. Few meat eaters are prepared to hear that their food decisions are somehow unethical, harmful, and costly to the environment. Like other destructive behavior, the meat habit is embedded in complex social relations and to some degree in ideological beliefs. A widely ballyhooed nutritious food, meat has long signified robust health while simpler foods (grains, vegetables, fruits) were often associated with inferior, cheap, bland diets of the poor and lower classes. Even today meat (above all, beef) represents *power*, especially masculine power, like that wielded by strong leaders and warriors, a kind of special nourishment needed to carry out tough work and athletic performance. Carol Adams notes that "According to the mythology of patriarchal culture, meat promotes strength; the attributes of masculinity are achieved through eating these masculine foods."[45] Indeed the meat-centered diet is still regarded as a source of great virility. With the planet rapidly moving past its ecological limits, and with destructive and resource-draining meat consumption on the global upswing, humans remain locked in a closed universe of myths and addictions, immobilized by years of inbred practice, sustained by high levels of corporate propaganda and marketing.

In the United States, the insatiable appetite for expensive meats, dairy products, and fast food has given rise to an unprecedented health crisis. Americans spend at least double per capita on medicine relative to other industrialized societies yet have some of the worst health indicators in the world—ranking thirty-seventh on overall indicators, according to one UN survey. The incidence of cancer, heart disease, diabetes, and other chronic afflictions stands at or near the top, with no less than 52 percent of the population considered obese. The long and expensive "war on cancer" has been, with a few achievements here and there, a dismal failure.[46] The widespread use of harmful pharmaceutical drugs, which produce adverse reactions killing as many as 200,000 people yearly, has become a commonly accepted American way of life.[47] Cancer, which kills nearly 600,000 people on average each year, can be understood as a defining, morbid symptom of food addictions, ecological crisis, and social decay in advanced capitalism. Meanwhile, meat consumption and disease have grown so intertwined as to end up almost invisibly connected.

In his justly praised *The China Study,* Colin Campbell organized comprehensive research linking diet and health across different regions of China, showing through massive statistical data the central role that meat-based diets play in cancer and heart disease.[48] While the empirical basis of Campbell's study is mostly limited to China, an oft-neglected part of the book is the connection established between corporate interests (food, drugs, medicine) and declining health indicators in the United States and elsewhere. Campbell devotes several chapters to the key influence of international giants like Kraft, Nestle, Johnson and Johnson, and Pfizer, which, through vigorous advertising, marketing, and lobbying campaigns, perpetuate a global apparatus of waste and destruction. Worse yet, these industries have managed to colonize government, most successfully in the United States, Campbell writing, "There is nothing better the government could do that would prevent more pain and suffering in this country than telling Americans unequivocally to eat less animal products...and more whole plant-based foods. It is a message soundly based on the breadth and depth of scientific evidence.... But instead of doing this the government is saying that animal products, dairy, meat, refined sugar and fat in your diet are good for you.... The covenant of trust between the U.S. government and the American citizen has been broken."[49]

Academic work on health and medicine, for its part, is largely subsidized by these same corporate interests. The vast majority of agencies, panels, conferences, and research projects are initiated and/or dominated by agents of industry. The highly respected National Institutes of Health (NIH), with its annual budget exceeding $30 billion, funds many centers of research and study, but precious little work is devoted to disease prevention, or to unraveling the linkage between diet and illness.[50] As Campbell notes, "the entire system of developing public nutrition information...has been invaded and coopted by industry sources that have the interest and resources to do so. They run the show."[51] Needless to say, one will look in vain for mainstream reports associating chronic diseases with meat, dairy products, and fast foods; the tendency instead is to discredit such linkages. Modern technocratic discourse in the medical field typically dismisses the mountain of evidence behind the superiority of plant-based diets for the well-being of humans, animals, and the global environment. Meanwhile, billions of people around the world aspire toward American-style food consumption and lifestyles—a major source of developmental unsustainability and a recipe for hastened ecological disaster.

The neoliberal mode of globalization has turned into a mounting disaster for food production and distribution, rooted as it is in centralized agribusiness, the commodification of foodstuffs, a heavy emphasis on fossil fuels, a dysfunctional meat-based agriculture, and a lack of community control over natural resources. This model, while seemingly efficient across the twentieth century, has become an unsustainable disaster, its legacy massive hunger to go along with spreading poverty and pollution. The transformation of corporate-based agriculture into a system more sustainable, human-scale, and democratic lies at the center of any future ecological politics.

Global Crisis and Resource Wars

No in-depth look at the global crisis is possible without addressing the growing threat of militarism and warfare on a world scale—a threat best understood within the context of neoliberal globalization, corporate expansion, shrinking resources, and increased blowback against American imperial power. Going back to World War II, if not earlier, U.S. pursuit of international hegemony has revolved around its grand economic and geopolitical ambitions, resulting in the most powerful empire the world has ever seen. The American imperial machine, with its sprawling war economy and security-state, has emerged as both instrument and benefactor of seemingly limitless resource accumulation. As the planet faces heightened economic contradictions and environmental crisis, intensifying conflict over natural resources (above all oil and water, but also land and crucial metals) emerges as a driving force of world politics.

Imperial wars have been part of American history from the outset: more than two centuries of armed combat against Indians, the large-scale importation of Africans for slavery, the colonial thrust Westward, the Spanish-American War, the twentieth-century expansion into Central America and the Pacific, and the establishment of a warfare-state behemoth during and after World War II. Empire became integral to American public life—government, the economy, culture, social arrangements, the media, and of course foreign policy.[52] The warfare state flourished in the context of perpetual military conflict (and preparation for conflict) from Korea to Vietnam to Central America, the Balkans, Iraq, and Afghanistan, with no end in sight. By the end of the twentieth century U.S. global power had become anchored to a worldwide network of bases, economic institutions, nuclear forces, surveillance, and intelligence operations, driven by a quest for natural resources, cheap labor, new markets, and geopolitical advantage. Military power depends not only on massive arms spending but an authoritarian imperial presidency, bipartisan ideological consensus, corporate partnerships, and a media culture that celebrates every U.S. intervention abroad as necessary to promote democracy, peace, human rights, and national security. Superpatriotism instills blind popular loyalty behind any foreign venture, reinforced by an American exceptionalism celebrated through the virtues of U.S. economic, technological, and military supremacy. Any opposition to U.S. global power, domestic or foreign, must be according to this worldview the work of diabolical enemies, villains such as fascists, Communists, terrorists, rogue dictators, warlords, drug traffickers, and so forth. Images of such demonized targets, dutifully portrayed as such in the corporate media, intersect with an ensemble of racial, cultural, and political stereotypes that Americans readily embrace in a frightening world of chaos and violence. Playing off such stereotypes inevitably gives the power elite wide flexibility to pursue superpower objectives.

Without abundant and relatively cheap natural resources, industrial growth everywhere would enter a rapid downward spiral. During the modern era the world has benefited from vast seas of oil relatively easy to locate and extract, inexpensive to produce and distribute, and of course essential to consumers. As mentioned, however, the earth is now moving into an epoch of peak oil, the point at which production can no longer keep pace with escalating global demand. As energy resources grow ever more scarce, developmental schemes long reliant on cheap fossil fuels will no longer be sustainable. By 2030 nations of the world are expected to more than *double* present energy intakes, with autos and electricity in the lead, a consumption level reaching perhaps 140 million barrels of oil daily. But with gradually declining supplies, combined with a reluctance of corporate and government leaders to break the oil addiction, this expectation is nothing more than a recipe for utter catastrophe. Paul Roberts, in *The End of Oil,* foresees a "swift, chaotic shift in our energy economy [that] almost guarantees disruption, uncertainty, economic loss, even violence."[53] Such dystopic predictions do not even take into account three other crucial factors: population growth, the harm wrought by climate change, and wasteful meat-based agriculture.

No society is more habituated to the oil economy than the American, which devours fully one-fourth of total worldwide fossil-fuel reserves. By 2030, however, American oil fields will be largely depleted, accelerating a desperate search for energy sources abroad just when global supplies are shrinking and worldwide demand is skyrocketing. Of some 45 oil-producing nations, current estimates are that more than half will soon (by 2020) reach peak output.[54] American business formulas assume perpetual growth on the basis of readily abundant energy sources, with possibly 10 to 15 percent reliance on "green" technology. Despite glib promises of "clean" alternatives, significant departure from the hydrocarbon economy is not presently envisioned: "clean" coal is nearly as destructive as its dirty forerunner, nuclear sources will be exhausted or rendered politically objectionable, and such technologies as wind, thermal, and solar energies will be severely restricted as to where and how they can be employed. The conversion of tar sands or shale into oil will continue to be expensive, difficult, and limited in volume. Natural gas is expected to peak not long after petroleum.[55] No remedy is therefore conceivable in the absence of a dramatic shift away from existing growth patterns, energy priorities, investment strategies, and infrastructure development—that is, away from currently dysfunctional modes of production and consumption. But in the United States at least, fundamental alternatives are more or less taboo within political discourse and public policy, hardly surprising for a system so wedded to corporate agendas, unfettered economic growth, a war economy, and privatized consumption. In a period of soaring military budgets, intensified global arms trade, ongoing civic violence, warfare, and terrorism, this illusory path augurs further descent into a Hobbesian state of nature.

The harsh and unyielding reality is that two centuries of accelerated energy growth, dominated by a few leading industrial nations, will soon be coming to an end. Corporate and military elites, unfortunately, want no part of this depressing prospect: ecological priorities are scornfully dismissed, their embrace likely to poison any political career, Democrat or Republican, as the "consensus trance" is emphatically bipartisan. Reduced growth? Socialized investment to rebuild and expand the public infrastructure? A racheting down of suburban living arrangements? Movement away from meat-centered agriculture and fast foods? Dismantling of the war machine and its worldwide network of military bases? A constriction of Wall Street casino operations? None of these alternatives, however reasonable and even practical, have been seriously contemplated at the summits of American economic and political power. Such "utopian" scenarios are left for a few progressive think tanks to ponder, quite removed from the orbit of decision making.

Rarely mentioned is the fact that the United States had already reached peak oil in the early 1970s, following which it underwent a slow economic decline marked by deindustrialization, the rise of casino financial operations, increased dependence on foreign imports, elevated budgetary and trade deficits, and greater debt-driven consumerism. By 2000 the United States was importing the bulk of its industrial goods and nearly 70 percent of its oil. New oil field discoveries in the United States have dwindled to a tiny percentage of total (and still growing) needs. On the American predicament, Heinberg comments, "It is our reluctance as a species to undertake demand-side solutions—not merely our inability to find a suitable substitute for oil—that is leading us toward collapse."[56] Whether this is a "species" problem or more concretely a question of corporate and governmental power, Heinberg points to the futility of locating developmental solutions in the discovery of new oil reserves; with the onset of peak oil, that pursuit only delays the inescapable for perhaps several years. And the "inescapable" calls for nothing short of a new economic model—a theme pursued more fully in coming chapters. Corporate capitalism remains thoroughly reliant on carbon energy sources, while Americans continue to be fearful of change; "green alternatives" are typically framed as new technological and market initiatives, compatible with the status quo. Meanwhile, what might be called "oil politics" is now enmeshed in a planetary web of power, intrigue, competition, and even military conflict as industrialized nations maneuver to bolster their position while emerging powers (China, India, Indonesia, Brazil, Mexico) assert their own insatiable energy demands. In this context, and given more than two centuries of U.S. colonial history and military interventions, Washington figures to be at the center of future resource wars.

By World War II American leaders had already turned their geopolitical gaze on the Middle East, site of roughly two-thirds of global hydrocarbon reserves—surely crucial to any national power seeking world supremacy. It was not until the Soviet collapse that this U.S. ambition could be given its fullest expression: in the early 1990s Washington moved to expand its military presence in the Persian Gulf, Central Asia, the Caspian Sea area, and the Balkans, hoping to "remap" this vast region to shape the flow of resources valued at many trillions of dollars. The goal was not so much "access" to resources, but rather "strategic domination" giving the United States an upper hand over all challengers, including Russia and China. And strategic domination could only be guaranteed in the last instance by means of superior military force, including nuclear advantage. This meant that American energy demands could, at least for the time being, remain as insatiable as ever—an addiction converging perfectly with the geopolitical fixation on the Middle East. Bush's Iraq venture can be understood as a function of two linked foreign agendas—economic imperialism and geopolitical leverage (including support of Israel). Future global conflict is thus likely to revolve around resource-laden zones of the Middle East and Central Asia. In *Resource Wars*, Michael Klare writes "that conflict over oil will erupt in the years ahead is almost a foregone conclusion," although that observation (made in 2002) now seems rather prosaic in the wake of subsequent U.S. ventures.[57] As the realities of peak oil arrive, a war-making dynamic will be difficult to avoid, especially with the fast-rising carbon demands of countries like India, China, and Brazil while the United States maneuvers to retain its precarious 25 percent share of global energy sources. Meanwhile, given an expanded military presence in the Middle East—combined with unyielding support of Israel—Washington can anticipate decades of blowback bringing perpetual conflict with Arab and Muslim populations, scarcely a formula for peaceful global relations.

In this setting, rapidly increasing worldwide demand for oil during the 1990s was bound to generate new levels of international turbulence and conflict. Worth noting here is that many leading fossil-fuel consumers—India, China, Japan, South Korea, Brazil—have few reserves of their own while the United States, as mentioned, long ago reached peak oil and became a net importer. Although Washington had for decades looked to control Middle Eastern resources, it was the 1991 Gulf War that signaled the first major salvo of modern resource wars. The Iraqi invasion of Kuwait offered an easy pretext for U.S. military intervention, justified as a move to defend Kuwaiti national sovereignty. American strategic inroads at the time, however, remained partial; Saddam Hussein was left in power, resulting in strident demands to "finish the job" from a growing circle of neocons who looked toward "regime change" and a client Iraqi state. It remained for Bush and Cheney to realize this agenda, made easier in the aftermath of 9/11 and a decade of U.S.-led economic sanctions against Iraq, reinforced by bombings and covert operations to weaken the Hussein regime. The neocons engineered congressional passage of the 1998 Iraq Liberation Act, which called for regime change, giving Bush a free hand despite failure to achieve UN backing. American influence over vast Middle Eastern oil reserves would be enhanced after the 2003 invasion and occupation of Iraq.[58] The fact that this operation was illegal and justified to the public by a series of shameless lies speaks to the urgent priority of resource agendas at the summits of American power—and the role of the media in legitimating such agendas.

The familiar debate over whether the United States attacked Iraq for mainly geopolitical (Israel-related) or economic (resource) interests need not be resolved here: both dynamics would seem to have been at work. That Iraq contains as much as 15 percent of world oil reserves was hardly peripheral to the U.S. drive for regime change. Immediate attempts by U.S. occupying forces to seize and protect valuable petroleum fields and facilities—for example, at the sprawling Kirkuk oil complex—speak volumes about U.S. resource objectives. So too did later moves to privatize what had been nationalized energy assets under Hussein. Despite unforeseen obstacles, such as a tenacious popular insurgency, U.S. capacity to shape governance in Iraq allowed it to influence the flow of oil reserves, limiting access to such rivals as China and Russia. While President Obama in 2008 promised a relatively quick "exit" from Iraq, all indicators pointed toward a long-term U.S. military, political, and economic presence in the country: a string of huge armed forces bases, tens of thousands of troops, an extensive network of private military contractors, the world's largest American Embassy, and a client state with little authority to decide trade, investment, banking, and foreign policy. An "exit" seems even less plausible given Washington's fixation on such countries as Afghanistan, Pakistan, Egypt, and Iran, not to mention its alliance with Israel. Meanwhile, the predictable rise of Islamic militancy in the region and elsewhere—partly a creation of U.S. policymakers in the 1980s to fight global Communism, partly the result of blowback, partly arising from Arab/Muslim outrage over Palestinian oppression—helps legitimate the U.S. presence as necessary to the "war on terrorism." The epoch of oil warfare, with all its horrendous implications, seems to have been launched.

That much of what American planners had in mind for Iraq and Afghanistan has not worked smoothly, or in some ways backfired, hardly suggests lessened dedication to resource ambitions. U.S. aspirations for Middle East hegemony have not abated any more than its larger imperial goals; geopolitical conflict over energy sources, after all, is only in its embryonic stages. Fossil-fuel access, indeed control, remains central to American foreign policy, and the capacity to manage the flow of Middle Eastern oil supplies (with its possibly 800 billion barrels) is clearly an elite obsession. As world oil markets become tighter and more volatile, this obsession is likely to build, as reflected in a series of National Security Strategy documents during the second Bush administration. Iraq is estimated to possess more than 100 billion barrels of oil, though probably more than that once the nation's unexplored regions are fully surveyed. By comparison, the United States, Canada, and Mexico together possess no more than 80 billion barrels, with Russia and Venezuela each at roughly 70 billion barrels and the entire African continent at something less than that. During military occupation a key U.S. aim was to get most of Iraq's 80 oil fields running at close to peak levels, while hoping to avoid sabotage. Although U.S.-based energy firms did not (as of 2011) establish direct ownership of Iraqi oil production, the long-term expectation is that the United States, with its strong presence, will exert a decisive influence over the production and sale of the nation's resources.

Meanwhile, several giant U.S. companies—Lockheed-Martin, Chevron, Bechtel, and Halliburton among them—had by 2009 expanded operations in Iraq and other parts of the region. Other Western corporations wait on the sidelines, hopeful of investing in the oil economy if and when the political situation stabilizes. Antonia

Juhasz documents at great length the U.S. economic agenda for Iraq, commenting, "President Bush has expanded the policies of corporate globalization through the barrel of a gun."[59] In violation of international law, the American occupying regime set out to restructure the Iraqi economy, revamping the financial sector, investment policy, foreign trade, ownership laws, utilities, taxation, and the media. Multinational companies have gained ever-widening access to petroleum and other resources that under Hussein had been nationalized. Beyond Iraq, the United States had since the early 1990s built a ring of military bases, spanning from Kuwait to Georgia, Turkey, Afghanistan, Uzbekistan, and Tajikistan—backed by extensive naval deployments and air power in the region. The CIA and the NSA have stepped up U.S. intelligence and surveillance capabilities in the Middle East. Although Washington has stationed hundreds of thousands of field troops in Iraq and Afghanistan, its military power depends increasingly on technowar—satellite-based information gathering, new generations of bombs and missiles, and unmanned drone planes that can attack virtually anywhere.

Viewed globally, the problem is starkly obvious: sustained economic growth pursued by ruling elites in powerful nations, all dependent on a shrinking reservoir of natural resources, drives the world toward sharpening economic conflict, national competition, and potential warfare. Unfortunately, no industrial power is anxious to depart from the dysfunctional growth model, reflected in the foot-dragging at earth summits in Rio de Janeiro, Kyoto, Copenhagen, and Cancun. More troubling, there is nothing in American history, politics, or culture to indicate that serious rethinking of this model is likely: ecological crisis, even where acknowledged, is typically seen as a matter for "market" solutions and/or technological fixes. U.S. leaders have frequently stated their desire to preserve the "American way of life," whatever the risks and costs. To maintain this (non-negotiable) agenda, Washington must carry out an aggressive foreign policy, relying on military superiority to gain the upper hand in resource wars. An ensemble of factors is at work here—the spread of petroleum nationalism, the specter of peak oil, global warming, and new manifestations of blowback.

American economic decline, however calculated, hardly prefigures a retreat from its imperial reach or ambitions, as is often suggested. In fact, superpower weakening in production and finance could just as easily be translated into augmented military vigilance, spending, and interventions, as indeed has already occurred. U.S. military supremacy is not in immediate jeopardy: Pentagon expenditures, despite right-wing posturing about "deficits" and "government debt," had by 2011 soared past the one trillion dollar mark, in addition to special allocations for wars in Iraq, Afghanistan, Pakistan, and Libya. From a strictly military standpoint, therefore, a declining imperial power could become more rather than less menacing. U.S. leaders will continue to rely on some form of partnership, or "coalition building," with friendly nations, whether through NATO or other regional alliances. They will also face increasing challenges from rising countries that were formerly on the periphery of the world system.

Future U.S. strategic designs could also focus on China, a nation with stupendously high economic growth rates and its own insatiable demand for oil and kindred resources. A rising power on the verge of equaling or surpassing American industrial output, China will not yield deferentially in the realm of resource competition, nor will it be intimidated by Western powers. U.S. indebtedness to China, along with a growing reliance on finished Chinese industrial goods, is well known. China's close relations with such nations as Venezuela and Iran are viewed in Washington as a geopolitical threat, while U.S. hostility toward Iran has forced Iranian leadership eastward, toward China, whose leaders would be perfectly happy to gain special access to Iranian oil reserves. A nation of 1.3 billion people, China faces its own resource shortages without its own petroleum reserves, while its industrial power poses a challenge to U.S. global hegemony—a source of tension sure to be magnified over time. China produces 97 percent of all rare-earth metals essential to high-tech goods, and has shown a tendency to restrict exports in this area while hoping for leverage in other areas. The limits of U.S. power are visible in the dangers of overreach: Washington already has large-scale military deployments in the Middle East and Central Asia, has combat operations in Iraq, Afghanistan, Pakistan, and Libya, and threatens war against Iran. Any serious confrontation with China—or even nations like Iran and Venezuela—would stretch American capabilities far beyond manageable resources. Given the extreme rightward direction of American politics, however, such irrational courses of action cannot be dismissed.

The special U.S.-Israeli alliance dates from 1967 and remains durable insofar as it serves the robust geopolitical interests of both partners. While the American media regularly depicts Washington as a neutral "broker" between Israel and the Palestinians, in reality the United States has always been a full-scale backer of Israel—economically, militarily, and diplomatically—fully assisting in the denial of Palestinians' rights and the blockage of a sovereign

Palestinian state, in alignment with the domestic Israel Lobby. Collaboration on financial, military, intelligence, and other shared interests gives rise to anti-American blowback among both Arabs and Muslims. Israel operates as something of a sub-imperialist power for the United States in the Middle East, the partnership solidified by a joint monopoly of nuclear weapons in the region—something rarely discussed in Western media and political circles but central to tensions with Iran. The United States and Israel have both turned their sights on Iran, a nation occupying a critical landmass with oil reserves exceeded only by those of Saudi Arabia. Iran is geographically close to Iraq, Afghanistan, and Israel and has tight relations with Russia and China. Washington in fact has long viewed Iran as a major target, having in 1953 overthrown the country's first democratic government in favor of the Shah's brutal (but "modernizing") dictatorship. Since the early 1990s American neocons have focused on Iran as a threat to Israel, partly owing to its support for the Palestinians and Hezbollah in Lebanon, though Iran has managed a stable *modus vivendi* with all of its neighbors. Since 2006, U.S. and Israeli leaders, viewing the Iraqi situation as more or less "stabilized," have escalated threats against Iran, depicted as an evil "hub of terrorism" with grandiose nuclear plans. Iranian leaders are denounced as tyrannical and expansionary, akin to Nazi Germany—though no empirical record of modern-day Iranian military aggression exists. The reality is the opposite of what U.S.-Israeli propaganda claims: it is the Iranians who are encircled by vastly superior and hostile military forces, from the Persian Gulf to Iraq, Afghanistan, Turkey, and Central Asia, where the United States possesses deadly air, ground, naval, and space capabilities. It was Washington, moreover, that instigated Iraqi attacks on Iran in the early 1980s. Persistent warnings, on the heels of baseless economic sanctions, coming from both Israel and the United States not only reek of deceit and hypocrisy but pose new threats to world peace, a violation of the UN Charter prohibiting unprovoked military threats. Still, as of 2011, the cacophony of warnings and threats (with "all options on the table") remained persistent.

The deceit and hypocrisy have no boundaries: as Iran pursues its legal nuclear options within the NPT framework, the real nuclear weapons monopoly in the region, as noted, is possessed by Israel and the United States. There is no evidence that Iran (an NPT member) has a weapons project but it is widely known that Israel (an outlaw state that sternly rejects the NPT, with U.S. approval) possesses between 250 and 300 nuclear warheads, more than enough to destroy Iran and the entire region. Meanwhile, the United States continues to support rising nuclear powers India and Pakistan, also both opponents and violators of the NPT. As for the United States, it carries forward a long-standing "first strike" doctrine, lays out a variety of contingencies for nuclear warfare, fails to ratify an international test-ban treaty, and, most crucially, has invested during the Bush and Obama presidencies hundreds of billions of dollars for nuclear research and development—all the while lecturing others about the perils of nuclear proliferation. Warmongering against Iran was taken up by the U.S. Congress in 2009, its politicians overwhelmingly backing a resolution for economic sanctions, covert action, bombing campaigns, and (implicitly) invasion and regime change. Should warmongering move to the next level, as the Israel Lobby fervently desires, the results (impossible to predict or manage) could bring devastating warfare to the entire region and indeed heightened blowback across the globe, with attendant economic chaos possibly leading, in turn, to new and more horrific cycles of military conflict. Nuclear warfare (laid out as rational Pentagon "contingency") can hardly be discounted, especially since the United States will want to attack deeply hidden and remote Iranian targets. U.S. plans for yet another militaristic exercise in "regime change" against a Muslim country will surely be resisted by millions of dedicated, well-trained Iranians, a situation that could push the world to the brink of global warfare.

A Strangelovian move of this sort would be fully supported by the Israel Lobby, the bulk of U.S. corporate and Wall Street interests, the leadership of both parties, a majority of religious and academic elites, the military and intelligence establishment, and of course, a jingoistic, cheerleading mass media. The warfare-media spectacles that accompanied two wars against Iraq would be tame in comparison with this new superpatriotic extravaganza. That such an outlaw venture might be a function of resource wars—or simply joint American-Israeli strategy to destroy Iranian power—would scarcely be mentioned in media coverage or political discourse, nor would it be openly debated. Instead, the public would be treated to a litany of phony claims: regime change needed to fight tyranny, promote democracy, and address the issues of WMD and terrorism. The monstrous Iranian "threat" must be extirpated by American military force and heroism, whatever the material and human costs. Should Washington choose this route, it will have been facilitated by an atmosphere of ethnocentric belligerency, new waves of material sanctions, a demand for total Iranian capitulation, the demonization of Iranian leaders, the (unspoken) pull of resource competition, and the rightward swing of American politics. Meanwhile, faced with ominous threats and encircled by hostile military power, the Iranians might understandably be forced into a nuclear weapons project as a matter of deterrence. The actual Iranian "threat," therefore, lies in the potential for self-defense rather than a Nazi-style blitzkrieg of the region or missile attacks on Europe and/or Israel.

In the midst of the global crisis, with even the industrialized West fighting off economic collapse, U.S. war-making moves against Iran might seem irrational, but in a world of intensified resource conflict such moves have their own (perverted) logic. Once the demise of Soviet power gave U.S. leaders greater flexibility on the world scene, aggressive moves in the Middle East soon followed, starting with the first Gulf War. As one *Monthly Review* commentary puts it: "The war in Iraq is best viewed as an attempt to assert U.S. geopolitical control over the entire Persian Gulf and its oil—an objective that both political wings of the establishment support, and which is part of the larger aim of the restoration of a grand U.S. hegemony."[60] The Iran "crisis" can be understood as extending this imperial obsession. War against Iran, however, will have more severe consequences than military conflict in Iraq or Afghanistan, given the level of Iranian power and prospects for nuclear war. Some argue that U.S. quagmire in Iraq and Afghanistan will force imperial power into retreat, toward more restrained global ambitions; continued interventions, it is said, impose unacceptable risks and costs in a period of economic downturn. The overpowering logic of resource wars, however, dictates otherwise: the superpower will to dominate is limitless and insatiable, still backed by the deadliest military apparatus in history.

Global resource wars will be hastened by intensifying contradictions of a world capitalist system dependent on perpetual growth, corporate rule, and abundant resources. Those contradictions are sharpened by several interwoven factors: peak oil, global warming, infrastructural pressures, escalating social and environmental demands, all unlikely to be addressed within existing modes of production and consumption. A militarized world inevitably flows from these contradictions, as national and corporate maneuvering over scarce resources increases. International powers, including the United States, China, India, Europe, and Russia, have stepped up military spending, making the globe more vulnerable to conventional warfare (already widespread) as well as nuclear catastrophe, blowback, terrorism, and civic violence. By 2011 China had begun to challenge U.S. armed forces supremacy in the Pacific, with its renewed dedication to naval power, space technology, nuclear weaponry, and cyber warfare. At a point when China can no longer sustain its extremely high growth rates because of the worsening contradictions already noted, its military stance—like that of other competing powers—could turn increasingly belligerent. The reality of a worldwide orgy of growth coming to a deeply unsettling end will hit the major powers hard, further destabilizing the global system. In such a cauldron of world tensions and national geopolitical competition, the specter of nuclear warfare cannot be discounted—a specter all the more conceivable as the United States and Israel ratchet up their crusade against Iran. Here three distinct global crises—economic, ecological, and military—have become thoroughly and dangerously interconnected as global leaders continue along the route of endless growth, resource conflict, and ideological escapism.

U.S. global ambitions will no doubt face new challenges and obstacles, but these are likely to be met with heightened superpower resolve. The United States has fought "counterinsurgency" wars throughout its history, at the cost of tens of millions of lives and untold destruction to built environments and the natural habitat—a legacy still alive and relevant to the era of resource wars. The deadly cycle of militarism and terrorism, neoliberal globalization and resource wars, corporate expansion and economic misery has already generated a downward spiral toward global barbarism—that is, economic crisis, ecological devastation, military violence, spread of WMD, and political authoritarianism—unless countered by radical political intervention on a world scale, and soon.

Chapter 3

The Political Impasse

The deepening global crisis demands a fundamental, and rapid, shift in production, consumption, and lifestyles, overturning many decades of human assault on the natural habitat. The world capitalist system, driven by perpetual growth and resource exploitation, has bequeathed a Hobbesian legacy of decay, turbulence, conflict, and violence engulfing every nation and culture. A degraded ecology negates the capacity of humans to build positively and creatively toward the future. In this milieu, elites gravitate toward a variety of cheery scenarios—technological fixes, "market" solutions, "green" alternatives, spiritual renewal—without challenging business-as-usual, holding to the same corporate agendas that produced the crisis. However, at a time when (in Bill McKibben's words) "nature is finally pushing back," such hopefulness amounts to sheer delusion: catastrophe can be averted only through the adoption of a new developmental model, a new economics, a new *politics*.[1] The world now faces a stark and imminent choice—radical change or "barbarism." At present the ruling interests seem perfectly happy to follow their familiar destructive course—ever-expanding accumulation of wealth and power within a corporate system that concedes no limits to economic growth and resource appropriation. Even as the tipping point nears, the battle cry of business elites and their media propagandists grows ever-more strident: more "private" investment, more expansion, more fossil fuels, more deregulation, more resource availability. This is a power structure that has become so institutionalized, so embedded in routine, as to be entirely trapped in its own intransigent logic. In the United States, such political impasse can only be understood in historical and structural terms, played out against the backdrop of multinational corporate power, a massive lobby complex, Wall Street plutocracy, the permanent warfare state, a huge media-entertainment system, and the great electoral spectacle.

Corporate Colonization

The postwar era of organized state-capitalism has meant, for the United States, a steady growth of corporate, governmental, and military power, giving rise to the most concentrated and far-reaching system of rule in history. Beneath familiar myths of the free market, pluralism, and self-governance it takes little effort to identify well-established oligarchic, bureaucratic, and statist tendencies within modern capitalism, a power structure penetrating every corner of the globe and ruled by an elite first analyzed by C. Wright Mills, who in the 1950s identified a confluence of interconnected forces that were less developed than today: corporate expansion, the war economy, the national-security-state, and globalization.[2] For Mills, the power elite was intent on colonizing every arena of public life: the party system, legislatures, government agencies, the media, universities, and foreign policy. Countervailing forces (labor, popular movements, local communities) were expected to weaken, although Mills could not have foreseen the explosion of social movements in the 1960s and 1970s. Anticipating an upward trajectory of corporate power, Mills would have been little surprised by the later consolidation of big business, Wall Street, and the military-industrial complex, along with a widening gulf between rich and poor, mass disempowerment, and an atomized, depoliticized citizenry.

With the corporate system presently dominated by a few hundred giant enterprises and run by a few thousand owners and managers—in command of preponderant wealth and resources—the question arises as to whether even truncated democratic governance can survive the thrust of such inegalitarian and authoritarian pressures. Could democratic politics thrive where a population is subjected to gross disparity of income, resources, and status combined with lack of popular control in workplaces, government, and other spheres of public life? If American society is today ruled by a narrow oligarchical elite, as I argue throughout this book, then classical democratic ideals grounded in nineteenth-century liberalism would now seem illusory, a relic of civics textbooks, campaign speeches, and patriotic rituals. If so, hopes for sweeping reforms to meet challenges posed by the global crisis must be viewed as dismal.

While early capitalism gave rise to the commodification of social life, over time this logic came to engulf more of human activity, expanding from production and work to the realms of culture, politics, and daily existence. In *The Communist Manifesto,* Karl Marx wrote that "the bourgeoisie, during its rule of scarce one hundred years, has created more massive and more colossal productive forces than have all preceding generations together." The scope of this epic transformation meant a gradual enlarging of domination insofar as "the bourgeoisie cannot exist without constantly revolutionizing the instruments of production and thereby the relations of production and with them the whole relations of society."[3] Commodification of the world was defined less by free markets and democracy than by a system of integrated power in which state and capital, society and economy would progressively converge. The manufacturing (and financial) apparatus came to impact public life as a whole, giving rise to what Marx called the "fetishism of commodities." While capitalism often brought forward liberal ideals compatible with nascent democratic norms and practices, it simultaneously led to concentrated economic power in the hands of a narrow oligarchy, with capitalism and democracy entering into a perpetually conflicted legacy. Nowhere was this developmental pattern more visible than in the United States, where the Founders (a tiny stratum of white European men) presided over a system marked by a rising bourgeoisie, slavery, the removal and destruction of Indian populations, and limited suffrage. Marx concluded that the concentration of capital was destined to increase across history: beneath the surface of liberal-democratic forms, corporate elites would eventually tighten their material, structural, and ideological hold over society. They would be further driven to globalize the process of capital accumulation, not only for profits and growth but for economic mobility permitting greater leverage over labor as well as (most) national governments. As Marx put it: "The need for constantly-expanding markets for its products chases the bourgeoisie over the whole surface of the globe."[4] Later, it would be necessary to add the search for cheap labor and natural resources and then, ultimately, the pursuit of geopolitical supremacy.

By the twentieth century, and especially in the aftermath of the 1930s depression, a more efficiently organized capitalism based on augmented state ownership, planning, and public regulation would become an established fact of modern life—whether under fascism, social democracy, or Keynesian liberalism, as in the United States. A merger of corporate and governmental power, resulting in a variant of state capitalism, would be analyzed by Max Weber and Joseph Schumpeter, thinkers associated with Austro-Marxism, "elite theorists" such as Robert Michels, and the Frankfurt School. At this point

Marx's prophesy regarding capitalism leading to oligarchy had been confirmed, though not precisely as an outgrowth of the crisis tendencies Marx had likewise anticipated. For Weber, Schumpeter, and the others, the advent of rationalized capitalism meant *state* capitalism with expanded social allocations, deep government involvement in the economy, and regulation of class conflict. Reflecting on postwar America, Mills would carry this perspective further in his analysis of *militarized* state capitalism—a historic convergence of corporate, government, and military interests fitting the U.S. role as leading world superpower embedded in a bourgeoning war economy and security-state. For Mills, the liberal-democratic façade barely concealed an American power structure gaining new capacity to exercise its dominion nationally and globally. Thus, "As the institutional means of power and the means of communication that tie them together have become steadily more efficient, those now in command of them have come into command of instruments of rule quite unsurpassed in the history of mankind."[5] As for the economy, the system even in the 1950s was controlled by no more than a few hundred large corporations destined to expand (and globalize) over time.

What gave this power structure unique force and longevity was a set of extraordinary historical conditions, including the absence of feudalism and the presence of an expansive frontier, allowing capitalism to achieve special legitimacy in the United States. Thus, "The American elite entered modern history as a virtually unopposed bourgeoisie. No national bourgeoisie, before of since, has had such opportunities and advantages."[6] Mills' thesis points to a crucial element of American exceptionalism: such an integrated system would probably not have been sustained, much less strengthened and globalized, had it not rested on a compelling ensemble of myths, values, and traditions. The widely accepted belief in a uniquely democratic politics was always central to this history, even where that belief starkly conflicted with reality. Today, even as ordinary citizens grow alienated from the levers of governance, they commonly embrace this fiction as the corporate subversion of participatory norms occurs through a confluence of factors—globalization, technology, deregulation, an oligopolistic economy, media concentration, lobby power, the decisive role of money in politics, authoritarianism endemic to the war economy and the security-state. Meanwhile, elites atop the corporate-state accrue new maneuverability to advance their interests, facing only mildly troublesome obstacles along the way.

Since the early 1990s the world has seen the growth of a transnational super-elite able to control the global economy through its institutional position in finance, industry, and global communications. A new and more aggressive plutocracy, this stratum is obsessed with the smooth flow of capital allowing for greater leverage over labor, local communities, and the weaker national governments. David Rothkopf writes that this "superclass" comprises about 6,000 agenda-setting leaders, based mostly in the United States and a few other developed countries, their power derived from a hundred or so dominant corporations and banks. This stratum occupies the summits of power for which ideological consensus around maximum growth, deregulation, government austerity, and hostility to social reforms is axiomatic.[7] Globally, the 250 largest corporations account for some $14 trillion in yearly sales (as of 2011) while the 100 biggest financial centers manage over $43 trillion in capital resources, sufficient to control the economic, political, and cultural terrains. The super elite has enough power and wealth to shape key decisions on investment, finance, labor relations, the flow of goods and services, and general social policy—with few if any democratic mechanisms to counter those interests. In 2005 the combined sales of the top five international companies exceeded the GDP of all but seven nations, with such giants as Wal-Mart, ExxonMobil, British Petroleum, and Chevron able to amass unprecedented wealth. Of the largest 106 mega-firms, 91 were based in the United States or Europe, with vast economic reach—bolstered by the World Bank, IMF, and WTO. As Rothkopf notes, "Networking among the corporate elite can thus take a variety of forms. Working together, doing deals together, sitting on boards together, even attending gala events together—all these things help forge the networks that empower and define the superclass."[8] Giant energy firms naturally occupy a special niche in the world system, indispensable as the basis of economic development and military power: behemoths like Shell, BP, ExxonMobil, Chevron, and Haliburton anchor modern industry and trade. In the United States ties between big oil, Wall Street, government, and the military become more intimate, facilitated by state subsidies, tax breaks, leases, and infrastructural supports. Energy lobbies are uniquely powerful in the Beltway and state capitols, another indication of how corporate and state interests converge. The energy sector, moreover, is in the vanguard of fighting environmental reforms and measures to limit pollution and carbon emissions. At a time when energy resources are being internationalized, corporate power is managed by largely U.S. interests, Rothkopf noting, "Today, companies dominate the superclass and Americans dominate the leasers of those companies."[9] And those companies operate mostly impervious to any global or domestic popular accountability.

Viewed thusly, globalization refers to a phase of capitalist development ensuring oligopoly—with even fewer gigantic corporations controlling the flow of resources, investments, and trade—as well as financialization, where "development" increasingly revolves around casino operations, risky ventures, and commodity speculation. Transnationalization leads to a heightened fusion of economic and political power worldwide, the triumph of integration over competition, heightened class exploitation, and weakening of countermechanisms to the world system. Defended by neoliberal ideology as the basis of free markets and democracy, this system represents just the opposite—an assault on labor, communities, and local movements as well as on nature, while sharpening inequality to the point where the wealthiest 2 percent own more than 50 percent of global wealth. If crisis is indeed endemic to this highly unstable and dysfunctional order, it is deepened by contradictions simultaneously at work in the economy, politics, social life, and ecology.

Lobbies: The Subversion of Politics

It is no secret that lobbies, plying their influence across the political landscape, have become the cornerstone of big-business interests in the United States. The merging of government and corporations eviscerates the border separating "public" and "private" forms of power, as thousands of well-funded, tightly organized interest groups work behind the scenes, often removed from popular access or even awareness. Lobbies mobilize great resources—money, organization, social networks, expertise, media access—to shape legislation, win tax breaks for clients, get lucrative contracts, gain state subsidies and bailouts, block or weaken regulations, nullify reforms, and infiltrate public agencies. As their power expands, divisions between legislators and lobbyists shrink dramatically, Arianna Huffington writing, in *Third World America*, that "with this merging of state and private power, we're getting to the point where the only difference between senior congressional staffers and the lobbyists and influence launderers whose ranks they'll soon join is the size of their paychecks."[10] In 2010 more than 15,000 lobbyists were operating in Washington, D.C., along with tens of thousands in state capitols. The notion of a balanced, pluralistic system of interest groups competing on equal terrain to influence legislation is archaic: business groups representing banks, insurance, pharmaceuticals, agribusiness, energy, and the military dwarf in numbers and resources those advocating for labor, consumers, and local communities. The uniform interest of lobbies—to maximize corporate power and wealth—routinely prevails over those of what Domhoff calls the "liberal-labor coalition."[11]

A recent thorough critique of the lobby complex is Thomas Frank's *The Wrecking Crew*, which charts the deterioration of a political system penetrated by Beltway ultraconservatives working to subvert social reforms and give *carte blanche* to freewheeling corporate interests. Expanded lobby clout took off with the 1980s Reagan Revolution, which viewed the liberal state as a perversion of social values and the "market" an ultimate embodiment of human potential, an ethos that guided the second Bush presidency. For libertarians weaned on reputed laissez-faire principles, the market represents all that is organic and natural while the state is innately coercive and alienating, the enemy of freedom, democracy, and prosperity. As Frank argues, however, this archaic ideology amounts to little more than shameless cover for unregulated corporate power, the authoritarian and plutocratic elements of which are never acknowledged—a pseudo-libertarianism serving to legitimate elite rule. Market pretenses allow false populists like Tea Party activists to "pass off a patently pro-business political agenda as a noble bid for human freedom."[12] The success of lobbies in colonizing government agencies has long depended on the capacity of business interests to colonize large areas of the political process. Stupendous wealth that influences electoral outcomes, "privatization" of government functions, proliferation of right-wing think tanks and foundations, infiltration of government, and tightened corporate media ownership all reinforce oligarchic tendencies.

In recent decades Washington has witnessed a massive influx of lobbyists, mostly from the insurance, banking, pharmaceutical, military, agricultural, and media sectors. Usually with measurable success, these interests seek leverage over legislators from both parties and such agencies as the Federal Communications Commission (FCC), Food and Drug Agency (FDA), Environmental Protection Agency (EPA), Mineral Management Services (MMS), and National Institutes of Health (NIH), looking to weaken or neutralize federal and state regulations. In *The Wrecking Crew*, Frank describes how, in order to advance their economic goals, free marketeers strive to extinguish the last vestiges of the Keynesian social contract, an objective taken up in earnest by the Tea Party. While Frank dwells little on postwar trends toward bipartisanship, the "winger wonderland" he explores is almost equally inhabited by liberals, few of whom resist the entreaties of big business. Most liberals in fact subscribe to the same "free-market" mythology and benefit from the same corporate largesse. Frank ultimately concedes that "It would be nice if electing Democrats was all that was required to resuscitate the America that wingers flattened, but it will take more than that... [It will require] a reconstructive project of massive proportions."[13] For relations between lobbies and the Democrats, one need look no further than the close ties between the Obama administration and operatives from Big Pharma and Wall Street.[14]

Nowhere is the lobby complex more active, and more obstructionist, than on the environment. Reforms of the 1970s and 1980s, made possible by active social movements, have been challenged and sometimes overturned since 2000, under attack from polluting industries (coal, oil, chemicals, mining, utilities), their powerful lobbies, and think tanks like the Heritage Foundation, Cato Institute, and AEI. Ecological backlash is funded by such foundations as Coors, Olin, Scaife, Castle Rock, and Bradley not to mention the multi-billionaire Koch brothers—a campaign sold to the public as vital to free markets, growth, and jobs.[15] The aim is to free offending industries from EPA regulations like the Clean Air and Clean Water acts. Backlash ideology, given new life by the Tea Party, depicts environmentalism as a diabolical (Marxist, Communist) plot to destroy not only free markets and economic growth but Western civilization itself. As with the compromised Tennessee Valley Authority (TVA), long ago taken over by private interests, corporate strategy is to penetrate the EPA so that corporations can nullify or perhaps temper public intervention. Officials recruited from the very polluting industries they are appointed to monitor can undermine EPA regulatory functions, a stratagem perfected during the second Bush presidency. Leading EPA figures under Bush and Cheney entered government from such backgrounds as energy, chemicals, timber, utilities, and mining. The White House director of the Council on Environmental Quality was James Connaughtan, a former asbestos industry lawyer.

As the EPA sadly became what Robert Kennedy, Jr., calls a "country club for America's polluters," the agency charged with protecting American citizens from environmental harm largely ceased enforcing the law.[16] The government looked away as mining, agribusiness, chemical, and logging interests routinely flouted regulations on air, water, and soil pollution. The White House dropped scores of legal cases against polluting corporations, a sector that contributed generously to the 2000 and 2004 Bush campaigns. Gail Norton, Department of Interior head, with ties to Coors, Amoco, and the Chemical Manufacturers Association, adamantly opposed business regulations and (in Kennedy's words) "opened up our public treasure to industry plunder."[17] EPA budgets were slashed, scientific data obscured, regulations subverted, and legal actions stonewalled under intense lobby pressure. This "new orgy of industry plunder" led to a rollback of federal emission standards, a relaxation of wildlife protection measures, military exemptions, subsidies for polluters, and the sabotage of efforts to reverse global warming.[18] Meanwhile, the nuclear industry retained veto power over appointments to the Nuclear Regulatory Commission (NRC), which, according to Helen Caldicott, "now virtually represents the nuclear industry... [with] the public effectively eliminated from the nuclear industry's decision process."[19] Owing to aggressive lobbies, nuclear enterprises have for decades received generous state subsidies for every phase of their operations—some $13 billion, including tax breaks, in 2005 alone when Congress passed its corporate-friendly energy bill. The Price-Anderson Act, going back to 1957, still requires taxpayers to fund up to 98 percent of $600 billion in government insurance to protect against nuclear accidents.

A sacred yet often pilloried American institution, lobbies have long been active at all levels of government, the most powerful of them funded by big business. Media lobbies work indefatigably to shape news, information, and entertainment in film, TV, radio, cable, and print outlets; military and related lobbies exercise influence over U.S. foreign policy; financial lobbies, emboldened by Wall Street fortunes, have been especially active in banking and economic policy since the 1990s; and insurance lobbies, highly visible during the 2009–10 health-care debates, have enjoyed a virtual stranglehold over the medical sector. Pharmaceutical groups, operating on behalf of what is known as Big Pharma, have achieved nothing short of legendary status in the Beltway. In these and most other sectors, well-organized lobbies do everything in their power to stifle debate, discredit opponents, furnish misleading information, and, most crucially, infiltrate and colonize public agencies. Toward these ends they can mobilize resources far superior to those of more dispersed labor, consumer, and community groups.

As for Big Pharma, the situation has turned increasingly dark: fewer than a dozen companies dominate this lucrative market, a semi-cartel powerful enough to determine conditions of manufacture, sales, trade, pricing, and indeed testing. Pfizer, Merck, Roche, Johnson & Johnson, and Novartis, for example, are all nearly identical in their operations, all fixated on the huge American market where demand for drugs continues to soar, thanks to aggressive sales and advertising. Corporations depend on government patents for exclusive marketing rights and public sector absorption of R&D costs, as well as legislation protecting against foreign competition and price fluctuation. Big Pharma casts its influence over Congress, state legislatures, the medical profession, government agencies, political campaigns, the media, and even academia, generally able to push through its agendas with little trouble. Behind the interest group umbrella PhRMA, pharmaceutical companies work nonstop to help shape (and limit) Obama's health-care legislation. Donald Bartlett and James Steele, writing in *Critical Condition*, devote many pages to the corporate takeover of American medicine, showing how a wave of mergers, acquisitions, and consolidations has solidified an empire built on technocratic structures, deregulation, stock market maneuvers, and super profits.[20] Lobbies representing pharmaceuticals, insurance, Wall Street, and the medical sector are crucial to outcomes that favor business interests over public interests, resulting in health-care cutbacks, employee layoffs, deskilling of the work force, lowered wages, and a decline in the quality of health care.

Big Pharma operations perhaps best reveal how megacorporations colonize the public sector through electoral campaigns, federal and state legislation, agency decisions, court verdicts, and media influence. In recent years nearly every member of the U.S. Congress and every White House occupant has enjoyed a cozy relationship with pharmaceutical lobbies. The PhRMA group is the largest such umbrella operation in Washington, deploying nearly 150 lobbying enterprises staffed by hundreds of well-paid operatives. Drug lobbyists are recruited from the ranks of former Congress members. Big Pharma donates tens of millions of dollars in campaign funds, most going to Republicans: drug industry executives gave generously to the Bush presidential campaigns in 2000 and 2004, but also donated to the 2008 Obama campaign. Front groups like Citizens for Better Medicine, masquerading as grassroots organizations, facilitate long-standing Big Pharma objectives like fighting price controls and cheap imports.[21] There was nothing in the Obama reforms, moreover, to disturb Big Pharma interests, while Congress passed legislation weakening FDA capacity to protect consumers, resulting in increased millions of *reported* harmful drug reactions yearly in the United States. Sidney Wolfe and associates state that in their 33 years of monitoring the drug business "the current pro-industry attitude at the FDA is as bad and dangerous as it has ever been. In addition to record numbers of approvals for questionable drugs, the FDA enforcement over advertising has all but disappeared."[22]

The power of American lobbies to deliver outcomes favorable to corporate interests and harmful to the public has only grown over time: in 2009–10, the Chamber of Commerce began a large-scale campaign funded by record amounts of money raised from corporations and rich donors. By early 2010 the Chamber had recruited nearly six million people to help with lobbying and elections, bolstered by the January 2010 Supreme Court ruling endowing corporations with a free-speech right to lavishly spend money on candidates, overturning a century of laws that had limited such spending. Elevated interest group power comes at a time when the major political parties were losing ground to well-financed "outside" forces, meaning lobbies. The Chamber set up a network called Friends of the U.S. Chamber, inspired by efforts to oppose (then repeal) Obama's health-care reform. In 2009 the Chamber spent $144 million on lobbies, a 60 percent increase over 2008.[23] The Chamber and kindred groups can now obtain vast funding from wealthy contributors without serious disclosure worries. While the Chamber officially represents three million companies, internal documents show that it depends overwhelmingly on just a dozen or so behemoths. Ostensibly bipartisan, it has routinely supported Republicans: in 2008, for example, 86 percent of Chamber political contributions went to Republicans.[24] The dramatic gains of rabidly conservative House Republicans in the 2010 midterm elections owed much to the Chamber's generosity.

The corporate billions poured into electoral and legislative processes are, with a few noteworthy exceptions, entirely legal, part of normal business at the summits of power. Even where lobbyists' behavior might be deemed unethical, it has become so widespread and routine as to be largely impervious to scrutiny or prosecution. Although by 2009 some 20 ex-lobbyists (out of many thousands) had been convicted of crimes, most influence-peddling passed beneath the radar, seen as just another expression of everyday politics. Even the convicted Jack Abramoff loudly complained that he had done nothing wrong, that he was just doing his job like all the others. "I felt my job was to go out there and save the world," he said. "I thought it was immoral to take someone's money and not win for them."[25]

Perhaps the biggest casualty of the lobby complex, as noted, is the environment: urgently needed reforms have been delayed, weakened, or gutted. Few state legislatures have mobilized enough votes to address global warming, for example. Nationally, while the House did pass a modest climate change bill, in 2010, it quickly died in the Senate with little media or political commentary—far less than what was devoted, for example, to the later Casey Anthony trial. The Kerry-Lieberman initiative—friendly as it was to big business— vanished without public notice, thanks to energy, utility, and auto lobbies. Even as the crisis worsens, for Congress it is more of the same: stepped-up offshore drilling, public subsidies to Big Oil and the auto industry, investment in "clean" coal and gas, renewed attention to nuclear power, weakening regulation of polluters. For its part, the Obama administration backed away from its earlier promise to fight greenhouse emissions. Not only Congress, but agencies of the federal government are so influenced by the energy industry that they often stand aside while the energy giants merely self-regulate, as with the events leading up to and accompanying the great BP Gulf of Mexico disaster in 2010. Offshore oil permits were given out by the thousands, including for rigs at dangerous ocean depths. It was the MMS, riddled with corporate interests, that looked away as BP was given *carte blanche* to set up its ill-fated Deepwater Horizon rig in the Gulf. In 2011, intense lobby pressures had convinced Obama to support a massive seven-billion-dollar 2,000-mile pipeline to transport crude oil (mostly from tar sands) to the Gulf Coast from Alberta—a project filled with potential environmental horrors, including aquifer destruction in the Midwest, oil spills, land despoliation, and worsening levels of both air and water pollution. Sold as an effort to make the United States less dependent on foreign oil while bringing perhaps 300,000 new jobs, the pipeline garnered endorsement not only from the White House and the Congress but also the State Department and many labor groups, all rushing to approve before any environmental impact report.

The global crisis ultimately forces departure from the corporate-growth model, but neither Republicans nor Democrats are prepared for such a break. New scientific data by such bodies as the NOAA, showing that recent years are the hottest on record, is muddied by relentless corporate lobbying. When global warming was first being addressed by the Senate in 2010, the lobby juggernaut responded in full force: OpenSecrets.org reported that oil, gas, and utility interests spent nearly $500 million to sabotage legislation on carbon emissions. Groups like the American Petroleum Institute and the National Petroleum and Refiners Association spent tens of millions on advertising, as did the Chamber of Commerce, insisting that reforms would kill growth, prosperity, and jobs. ExxonMobil also contributed its millions. Not only Republicans but also Blue Dog (and kindred) Democrats, themselves beholden to polluting industries, were easily seduced by such claims. The fight was carried to the EPA, even as members of Congress looked to neutralize the agency's capacity to fight global warming through the Clean Air Act. The impasse drastically worsened after the 2010 midterm elections, as 86 percent of incoming Republicans vowed to oppose any climate change legislation while fully half enlisted in the ranks of global-warming deniers.

An especially aggressive lobby operation, financed in part by the Koch brothers and ExxonMobil, is the innocuous-sounding American Legislative Exchange Council (ALEC), perhaps best known for its crusade to "Warm Up to Global Warming." According to corporate-bought ALEC activists, trading in junk science, climate change is nothing but a propagandistic attack on "free markets" by the EPA and its supporters, who ignore crucial evidence showing that carbon emissions, far from being a pollutant, are actually a substance that "enriches the atmosphere." ALEC insists that carbon and other emissions should go unregulated as government intervention is based on phony liberal science that only serves to block growth and jobs. Waging its own war on the EPA, ALEC operatives have won over the vast majority of Republicans (along with some Democrats) to support jettisoning of the agency. In the meantime, ALEC has convinced most Republicans to back a two-year moratorium on *any* climate change legislation, eagerly taken up by many conservative think tanks, the Tea Party, and Republican candidates for the White House. With ALEC the vast American culture of global-warming denial has reached its apogee.

Despite some variations in political style and delivery, the Obama administration by late 2011 had departed little from Bush on key environmental issues. Obama decided not to push for tough global-warming measures nor far-reaching changes in energy use, aside from an obligatory focus on green technology; corporate lobby power generally held sway. Obama seemed to recognize the climate change battle as unwinnable, which elicited anger from even liberal reform groups like Al Gore's Alliance for Climate Protection. Given the larger global power matrix, positive steps were unlikely to emerge from international meetings at Copenhagen (2009) and Cancun (2010): U.S. obstructionism was enough to block genuine moves to limit carbon emissions, though Washington was joined in this by others, including China and Russia. If Copenhagen ended in stalemate, never going beyond voluntary standards, the results at Cancun were no more encouraging as binding national targets were rejected despite a more amicable gathering; global consensus remained a distant vision. The notion of Obama as dedicated liberal was revealed as just another myth as Democrats, attached to the same energy lobbies as Republicans, had long before emerged as a party of capitulation.[26] Nebulous references to "change" and "hope" barely concealed a politics of "centrism," nowhere more so than on the environment.

More problematic yet, environmental groups too had mostly succumbed to corporate pressures, many retreating from earlier progressive stances on global warming.[27] Dependent on big-business largesse, such organizations as the Environmental Defense Fund, Sierra Club, World Wildlife Federation, and Nature Conservancy were reduced to paying lip service to the idea of reforms, fearful of alienating big-money sources and mainstream opinion. Johann Hari—writing in the *Nation*—notes, "They take the money, and in turn they offer praise, even when the money comes from companies causing environmental devastation."[28] Some group leaders even questioned the consensual science behind climate change, with others arguing that legislation is "unwinnable"—meaning anything opposed by Republicans, Blue Dog Democrats, and corporate interests. The largest of all environmental groups, the Sierra Club, suffered a precipitous drop in membership (from 714,000 in 2005 to 616,000 in 2011) owing in great measure to its overwhelming focus on large donors and its increasing coziness to wealthy contributors. As the organization moved closer to big corporations and utilities, restive members demanded change, leading to the resignation of Carl Pope as Sierra Club leader in late 2011.[29] With massive funding from oil giants like Shell, BP, and ExxonMobil, helped along by the Chamber of Commerce, environmentalism in the United States had arrived at a tragic impasse just when the ecological tipping point was moving ever closer.

Wall Street Plutocracy

As finance capital gains renewed power in American society, the banking system, Wall Street, and Federal Reserve coincidentally expand their political influence. In partnership with the World Bank, IMF, and WTO, and bolstered by economic globalization, these institutions work tirelessly behind financial leverage, open capital flows, and maximum freedom from legal and political constraints. The financialization of capital, long a conservative agenda, marks a triumph of capitalism over democracy, oligarchy over public interests, Wall Street over Main Street. The question here is whether the environment, or indeed any progressive social value, can survive this trend toward unfettered neoliberal corporate and financial order.

The great U.S. financial meltdown that began in 2008 represents a crisis more fundamental than earlier cyclical downturns, as it is the outgrowth of a newer phase of finance capital in which the system, the province of fewer corporate giants, is subjected to more extreme centrifugal tendencies than in the past. Despite all the celebration of "free markets," limited government, and individual responsibility, new efforts to stabilize Wall Street have required a raid on the public treasury assisted by the Federal Reserve, with taxpayers footing the bill for much of the larcenous gamblers' debts. President Obama, like other occupants of the White House reliant on Wall Street financiers, made several trillion dollars available to these interests in the form of cash infusions, loans, guarantees, and buyouts—all pushed strongly by bankers, the Fed, an army of lobbyists, and the media even as Obama was savaged by right-wingers as an out-of-control "socialist," using state power to take over American society. Meanwhile, the number of large banks dwindled to a mighty five—Citigroup, JPMorgan Chase, Bank of America, Wells Fargo, Morgan Stanley—all recipients of government aid as they purchase or drive out hundreds of smaller banks. A tiny group of financial giants held nearly 60 percent of total American banking assets in 2011. The Bush and Obama "stimulus" on behalf of these financial empires scarcely reached ordinary Americans needing their own bailouts, jobs, and public services. While top executives received tens of billions in pay, bonuses, and stock options, the "real" Main Street economy was deteriorating with lost jobs, home foreclosures, eroded infrastructure, declining personal income, and shrinking retirement and savings accounts. For its part, the Fed continued to pump cheap money into the financial system as perpetual stimulus to banks, while speculation continued, largely detached from the production of useful goods and services—and undeterred by the toothless Obama Wall Street reforms of 2010.

As the financial crisis deepened, banks stepped up their tenacious lobbying, ever hopeful of bailouts but, more crucially, fearful of regulatory pressures that could result from a groundswell of public anger. The eight largest banks ramped up lobby spending by 12 percent in 2009, to roughly $30 million.[30] JPMorgan Chase led with $6.2 million for 30 operatives in Washington alone. The Financial Services Roundtable, representing more than 100 firms, redoubled its efforts to kill or water down possible banking regulations such as size limits or caps on hedge funds, viewed by the Roundtable as "political interference." Obama's modest (though overdue) Consumer Finance Protection Agency ended up with so many loopholes and exemptions, not to mention a leadership void, as to be essentially powerless. This was inevitable given that the White House economic team was recruited mostly from the ranks of Wall Street insiders like Robert Rubin, Larry Summers, Timothy Geithner, and Michael Froman.

The entire political system, from the White House to the Congress to the Federal Reserve, is riddled with financial lobbies that rank among the most powerful in the Beltway and state capitols. In 1999, thanks to Wall Street pressure, Congress passed the Financial Services Modernization Act—followed quickly by the Commodity Futures Modernization Act of 2000—which tore down the 1933 Glass-Seagall Act that had erected a divide between commercial and investment banking. After 2000 Wall Street took on the *modus operandi* of a vast casino enterprise, investing funds largely unrelated to actual production or human welfare. A deregulation wave was spurred by the aforementioned army of lobbyists: real estate and mortgage brokers, insurance firms, credit card enterprises, investment banks, trade groups, private equity companies, hedge fund operators, credit unions. This trend built on a deregulatory frenzy going back to the early 1980s, when Wall Street began sending "oceans of money" into the governing arena. In Congress, major banks achieved virtually everything they sought, as the vast majority of politicians depend on financial support to get elected. Senator Charles Schumer (D-NY), for example, raises hundreds of millions from Wall Street for Democratic politicians while having received $14 million himself (in 2010) from the banking industry.[31] The symbiotic relationship between Wall Street and government has grown cozier in recent years, to the great detriment of public welfare and social reforms. Much legislation, as with the aforementioned 1999 and 2000 bills, is in fact drafted, edited, and revised by lobby operatives.

At the top of this financial pyramid is the sprawling Federal Reserve Bank, which has steadily accumulated power over the economy since its founding in 1913. Indeed the Fed epitomizes the historic convergence of government and big business, public sector and private banking, serving as the "lender of last resort," an abundant source of money supply, a regulator of interest rates, and the theoretical guarantor of fiscal and monetary stability. It supposedly constitutes a buffer against crisis, a moderator of extreme capitalist swings and cycles. The Board of Governors, in charge of the largest financial empire on the planet, is appointed to staggered 14-year terms, deliberately removed from any democratic inputs. The Fed chairman is designated by the president to govern an ostensibly "independent" body. Members of the Board, along with those of 12 regional boards, are recruited mainly from the banking industry. With no reliance on congressional funding—it earns money from Treasury securities—the Fed is essentially free of political oversight and can operate on its own without prior approval from the Congress or the president. The Fed also manages close relations with many foreign banks, typically outside any system of public monitoring or accountability. As one sign of its enormous clout, in March 2009 the Fed owned $247 billion in gold and held $534 billion in national debt.

The Fed has augmented its capacity to shape economic policy since the 1970s. It is among the largest, most self-sustaining bureaucracies in the United States, though without elected leadership. With Alan Greenspan as chair between 1987 and 2006, the Fed reached nearly mythical status for its power to steer the economy, as Board officials typically worked in secrecy with minimum oversight. As Robert Auerbach shows in *Deception and Abuse*, the Fed can manipulate financial markets on its own, provides no transcripts of its meetings, and routinely stonewalls even friendly congressional inquiries.[32] The Fed has enough resources to work independently of the three branches of government, its main constituency remaining banks and corporations. It is indispensable to oligarchic power, all the more so given its image as neutral crisis manager above the realm of mundane politics. Treasury Secretary Geithner remarked in 2009 that the Fed "defends the freedom and security of Americans from existential threats." Ben Bernanke, appointed chair in 2006, carried forward Greenspan's pro-corporate, easy-money policies, for which he was called "Bailout Ben, the patron saint of Wall Streets greedheads" and "King Ben, the unelected Czar of the fourth branch of government."[33] As they wrote this about Bernanke, *Time* magazine editors had chosen the Fed chief as "Person of the Year" for 2009 because "he is the most important player guiding the world's most important economy."[34] The problem with this accolade was that Bernanke's Fed, immersed in Wall Street larceny, itself facilitated the crisis as the Fed extended Greenspan's laissez-faire approach to the casino economy, issuing easy money and fighting regulations at every turn. The magazine's glorification of the Fed chair as autocratic savior of the economy unwittingly pointed toward the unchallengeable powers at the summit of a profoundly undemocratic system.

What Robert Scheer calls "The Great American Stickup" involved a feeding frenzy among a small, insular group of millionaires, who, driven by insatiable and reckless greed, pushed the economy toward its deepest crisis since the 1930s.[35] The March 2008 Bear-Stearns collapse triggered a series of crashes leading to bankruptcies, job losses, debt increases, and consumer paralysis that would exert a strong ripple effect across the global landscape. The havoc caused by speculative investment practices and banking schemes led to millions of home foreclosures, job losses, personal bankruptcies, and fiscal crises in state and municipal governments to be felt for years if not decades. Causes were traceable back to Reagan-era deregulations, tax cuts for the wealthy, huge increases in military spending, special funding for wars, and a dramatic rush toward finance capital—resulting, among other things, in the shipping of millions of industrial jobs abroad. The ascendancy of casino-style capitalism has ideological origins in the anti-Keynesian turn beginning in the 1980s. Congress, flooded with hundreds of banking lobbyists, has done little to restrict the freedom of Wall Street to continue business-as-usual: reforms did not curb high-risk trading or scale back the power of Too-Big-To-Fail banking giants. Strong Republican gains in the 2010 midterm elections emboldened new lobby efforts to subvert the most limited Democratic initiatives. As expected, groups representing Citibank, Morgan Stanley, Goldman Sachs, and JPMorgan Chase, among others, mobilized to resist measures that might interfere with Wall Street thievery.

A report issued by the Financial Crisis Inquiry Commission in January 2011, based on some 700 interviews over an 18-month period, identified a "gold rush" mentality in the American economy leading to the meltdown. The panel, comprising a Democratic majority, targeted executives at such firms as Goldman Sachs and Countrywide Financial Group for reckless investments and officials at the Fed for allowing this to happen. Commission chair Phil Angelides said, "The captains of finance and the public stewards of our financial system ignored warnings and, importantly, failed to question, to understand, and to manage the evolving risks in the financial system that is so essential to the well-being of our country."[36] Fed chiefs Greenspan and Bernanke, along with Treasury Secretary Geithner, came under withering attacks for failing to act effectively before and during the crisis. Republicans, for their part, sought to deflect attention away from Wall Street toward forces at work in the "global economy" and the public sector, where "fiscal crisis" would become the urgent order of the day.

As of 2011, the financial saga continues, with no end in sight. Operatives from Goldman Sachs and other firms have entered volatile markets such as global food commodities on a large scale, part of a U.S.-driven international speculation frenzy. By means of well-calculated strategic buying and selling of such foods as wheat, rice, and corn bankers can leverage huge flows of capital, driving up food prices while jeopardizing supplies around the world—a process endemic to a new era of investments in commodities, including foodstuffs. Global markets are perilously transformed, made more volatile. According to Frederick Kaufman, writing in *Harper's*, "Bankers had taken control of the world's food, money chased money, and a billion people went hungry."[37] Thanks to speculative markets, food prices worldwide had risen by 80 percent between 2005 and 2008 before at least temporarily settling down. Such a massive flow of capital into food commodities only portends future disaster at a time when local populations are increasingly disempowered.

Despite the financial chaos at the core of the historic economic downturn, Wall Street operations continue unabated, with few restrictions on casino ventures, no curbs on massive executive salaries and bonuses, no closure on Too-Big-To-Fail banks, indeed no effective consumer protection or public accountability. In a word, jungle capitalism has been allowed to thrive, its financial power in the hands of a few, as both parties shamelessly defer to the mania of deregulation and the myth of corporate responsibility. One of the major culprits, Goldman Sachs alone hired no less than 14 aggressive lobby firms to defend their (entirely unsustainable) power. Meanwhile, House Republicans and their Tea Party backers, fearing an Obama "socialist takeover" of Wall Street, pushed for even *less* government regulation of a financial sector still hell-bent on its global schemes, massive buyouts, and risky derivative investments, its managers commanding tens of billions in salaries and bonuses while having received hundreds of billions in public bailouts.

The expansion of (loosely regulated) financial power in the United States illuminates the historical conflict between capitalism and democracy, "private" interests and public welfare, "growth" and the environment. Elite power wins new autonomy in a matrix of reckless investment practices, recurrent bailouts and "stimulus" packages, scandalous lobby practices, and a media all too willing to ignore or finesse the true nature of the financial colossus. The recent crisis, revolving around government deficits and calls for severe public cutbacks, created a political milieu in which reform prospects (anything costing money) are effectively checkmated. As deficit hawks take over new ground on the terrain of ideological discourse, talk of new programs, including urgent "green" measures, becomes irrational since resources are simply nowhere to be found. As the White House, Congress, Fed, and media proclaim fealty to "main street" and "free markets," therefore, elite loyalties stray little from the Wall Street plutocracy whose interests directly counter those of ordinary citizens facing job losses, home foreclosures, pay decreases, crumbling social programs, environmental disaster, and a threatening future.

The Warfare State

With the postwar ascendancy of U.S. global power, the war economy and the security-state have gained a dominant presence in American society, the military becoming central to national identity, a symbol of power and supremacy most Americans take for granted. Foreign policy consensus behind superpower agendas is so absolute that even mild criticisms of U.S. global behavior—or massive Pentagon expenditures—seem offensive, revealing how the warfare state is now firmly embedded in the social and political order. Andrew Bacevich is correct to note that "the citizens of the United States have essentially forfeited any capacity to ask first-order questions about the fundamentals of national security policy."[38] The security-state was erected in the aftermath of World War II and with the onset of the Cold War, crucial to the new era of superpower ambitions. Organized around several intelligence agencies, the National Security State involved a wide range of diverse activities: surveillance, field investigations, high-tech communications, covert activities, espionage, law enforcement, and recurrent armed interventions. It currently possesses all the authoritarian features of a militarized corporate-state that undermines popular governance in the name of "national security." The closed universe of foreign policy narrows political debate, with exchanges of views the special domain of "experts" (usually conservative white men) at the top of the power pyramid. Citizen participation here threatens the smooth operation of the NSS, which requires maximum flexibility, speed, and often secrecy in a world of menacing enemies.

The warfare state rests on a complex ensemble of requisites—the capacity to militarily intervene abroad, massive Pentagon spending, high-tech combat, nuclear supremacy, bolstered executive power, a weakened Congress, and diminished checks and balances. It demands far-reaching surveillance and intelligence functions. It solidifies (and justifies) the familiar imperial presidency that, as President Truman observed approvingly after World War II, would allow the United States to freely pursue its national interests anywhere on the globe. In *War Powers*, Peter Irons writes that "never before has the planet faced a worldwide 'marshall' with such a massive arsenal at its disposal and with no institution, domestic or international, willing or able to restrain him."[39] The theme of Irons' book is summed up by his observation that "in a very real and very dangerous sense, the imperial presidency has hijacked the Constitution to serve the interests of the American empire."[40] The extent to which the Constitution was actually "hijacked"—as opposed to being made largely irrelevant—is questionable, but Irons is correct to note the trend toward a more authoritarian presidency. The postwar spread of U.S. global power through a network of financial and corporate structures, client states, military bases, surveillance operations, covert actions, and diplomatic venues indeed gives rise to a system of undemocratic institutions, practices, and norms. The time-worn military tenets of patriotism, loyalty, discipline, and violence extend beyond the Pentagon itself, penetrating ever-larger zones of American economic, social, and political life.

Neither the security-state nor the imperial presidency was an invention of the second Bush administration, as is commonly assumed. The concept of "unitary" executive powers was already present in the Theodore Roosevelt administration, and later gained ascendancy with the Cold War and the formation of the NSS, when the projected threat of Communism lent itself to heightened executive flexibility to confront new global challenges. "Crises" meant that lengthy and possibly conflicted public deliberations over matters of war and peace could threaten "national security." Truman was the first president to fully seize on independent or "inherent" executive powers, about which the Constitution was basically silent. Whatever limits democratic governance might have imposed on executive power—by means of checks and balances or advice and consent—would be overridden in practice, although honored in theory. Postwar U.S. history accordingly witnessed the presidential "takeover" of foreign policy, the work of Democrats no less than Republicans; military operations could be undertaken at the whim of the White House, with Congress assigned the role of writing checks: Truman in Korea, JFK in Vietnam, Nixon in Laos and Cambodia, Reagan in Central America, the first Bush in Panama and Iraq, Clinton in Somalia and the Balkans, the second Bush in Afghanistan and Iraq, Obama following Bush's agenda while adding Libya. Routine activities (war, covert operations, surveillance, proxy interventions, etc.) were undertaken in secrecy or, in most cases, automatically endorsed by compliant members of Congress. Some international agreements—between the United States and Israel, for example—were covered by "executive agreement," with congressional input bypassed or rubber-stamped. U.S. presidents nowadays typically approach Congress as a weak, indecisive obstacle to be easily finessed: the legislative branch might be "consulted" or "informed," but rarely invited as full decision-making partner. Even the widely praised War Powers Act, passed in 1971 to restore congressional "consent" following the Vietnam War, has been routinely sidestepped by imperial presidents. With politicians deeply in debt to the military-industrial complex, or cowed by patriotic appeals, as Chalmers Johnson notes in *Nemesis*, Congress does nothing on its own to check imperial war powers.[41]

The Bush-Cheney White House took the NSS to new heights, adopting strong "unitary" powers, preemptive military strategy, enhanced secrecy, expanded surveillance, and the license to sidestep or violate legal norms such as prohibitions against torture and warrantless wiretapping. The 9/11 attacks provided new ideological rationale for such initiatives, Bush anointing himself as "war president" who ought to be relatively free from public constraints on executive power. Charlie Savage observes, "The object wasn't just to strengthen President Bush's powers personally, but rather to strengthen the office institutionally, for all future presidents of both parties."[42] The war on terrorism could justify most any executive action once it met the dictates of "national security" or defense against foreign villains. New memos were drawn up at the attorney general's office establishing legal arguments for "extraordinary rendition" by the military and the CIA. The warfare state acquired new legitimacy in the face of projected new threats, allowing for increased military surveillance, intelligence, and law enforcement. The Homeland Security Office was set up in November 2001 to coordinate some 40 government agencies, including the CIA, NSA, FBI, and INS—a structure commanding more resources and institutional reach but less accountability. The USA Patriot Act was passed in October 2001 with little public input or congressional deliberation, the 342-page document giving the federal government sweeping powers to investigate and monitor electronic communications, personal and financial records, computer files, and related materials. By 2002 the war on terrorism brought new levels of governmental control and surveillance that later somehow escaped the notice of Tea Party protests against "big government."

In 2002 the Bush administration listed 153 terrorist organizations as threats to national security, enlarging Clinton's modest list compiled in the mid-1990s. Anyone belonging to such groups, however fleetingly, could now be interrogated, arrested, deported, or have their assets frozen with few legal protections. With the combined resources of the Departments of Treasury, State, Justice, and Defense—along with Homeland Security—the federal government was empowered to fight not only terrorism but also other challenges to the status quo: individuals and groups could now be targeted according to vague and arbitrary "terrorist" criteria. Local police agencies collaborated with the FBI, CIA, and INS to create their own intelligence units and antiterrorist squads, while a more expanded military role in law enforcement was hastened by terrorist fears combined with stepped-up immigration and drug operations along the U.S.-Mexico border. The protracted and costly war on drugs figures centrally in this saga, going back to President Nixon's early crusade and then, in the 1980s, to President Reagan's fixation on crime and drugs as twin evils to be eradicated as a cornerstone of national security. This "war" has been partially militarized, with especially harmful consequences for Mexico and Colombia.

Post-9/11 traumas helped revitalize an already swollen intelligence apparatus, weakened in the 1990s and later, its competence questioned because it failed to anticipate the terrorist attacks. The CIA, NSA, and related agencies now commanded more resources, had broader powers, and enjoyed stronger legitimacy. Meanwhile, the Pentagon and these other agencies became more integrated at a time when international and domestic intelligence functions merged. The capacity of this sophisticated network to invade personal freedoms and privacy on a large scale—thanks to new technological capabilities—cannot be overstated, as the Orwellian specter persists regardless of who occupies the White House. Further, as the 2010 Wikileaks revelations make clear, the NSS has long been active in gathering myriad types of global information. Washington directives have been uniformly sent to U.S. diplomatic personnel around the world, directing them to collect personal, technical, and political intelligence on counterparts and others at embassies, consulates, and other sites, including the United Nations, where American spying on members is a common occurrence.

As for the CIA, it has long served the White House as a largely private and unaccountable army engaged in covert operations. Its long record of international mayhem is by now too well known to merit recapitulation here. In *Nemesis* Johnson writes that "the CIA belongs as much to president as the Praetorian Guard once belonged to the Roman Emperors."[43] He adds, "In fact, the president's untrammeled control of the CIA is probably the single most extraordinary power the imperial presidency possesses—totally beyond the balance of powers intended to protect the United States from the rise of a tyrant."[44] While the CIA and kindred agencies enable the White House to conduct its activities through a massive web of secrecy, intrigue, and controls, the warfare state veers toward ideological closure, public input always checkmated by more urgent global priorities. As for Congress, few politicians dare question the work of this "Praetorian Guard."

Even more troublesome, while the American public is vaguely aware of such agencies as the CIA and the FBI, the far more powerful—and more intrusive—operations of the NSA proceed mainly in the dark, with even less oversight than given the CIA. Its electronic empire of nearly total surveillance remains impervious to public scrutiny. By 2009 the NSA had grown into the largest, most costly, and technologically developed intelligence agency in history, with over 200 million computers processing tens of millions of communications items literally every hour, linked to thousands of miles of fiber-optic cable and dozens of satellites circling the planet. The agency shares global information with several law enforcement and military organizations, part of the barely known National Security Operations Center (NSOC) created after 9/11. The NSA also has close working relations with military contractors like Raytheon and Lockheed Martin, as well as telecommunications giants like AT&T. Following 9/11, as the NSS moved toward full-spectrum surveillance, President Bush issued a secret order to step up warrantless wiretapping under NSA auspices. In summer 2007 Bush initiated crucial Foreign Intelligence Surveillance Act (FISA) reforms giving the NSA a freer hand, making it easier for the agency to monitor electronic transmissions, both domestic and international.[45]

NSA capacity to amass huge data warehouses gives it unmatched power to spy on Americans with little if any legal justification. An invisible fortress, the NSA remains as secretive today as when President Truman established it in 1949, without congressional approval or even public knowledge. According to James Bamford in *The Shadow Factory,* "The principal end product of all that data and all that processing is a list of names—the watch list—of people, both American and foreign, thought to pose a danger to the country."[46] The agency fiercely resists even minimum demands for accountability. Operating in an even more rarefied, insular, high-tech environment than other intelligence organizations, NSA directors rarely appear before Congress, the media, or the public. NSA relations with the telecommunications industry remain cozy and secret, allowing open access to the databases and technology of such corporations as Bell, Verizon, Microsoft, and AT&T. The Orwellian potential of this apparatus is indeed frightening: technology permits the NSA, working with corporations, the military, the police, and other intelligence agencies, to push the warfare state into a world of full-spectrum information awareness. Bamford shows how the NSA tirelessly pursues maximum surveillance to monitor phone calls, credit card receipts, social networks, GPS tracks, Internet transmissions, and online book purchases.[47]

In the area of federal law enforcement, the FBI has a long history of legal and political abuses going back to the early days of J. Edgar Hoover's leadership, worsening with the growth of the security-state. In 2010 the Electronic Frontier Foundation used the Freedom of Information Act to obtain hundreds of FBI documents revealing a pattern of violations between 2003 and 2008 alone, with at least 800 instances of *reported* violations, documents showing that FBI intelligence investigations have compromised the civil liberties of American citizens far more frequently, and to a greater extent, than was previously assumed. The FBI was discovered to have violated government policy and laws as many as 3,000 times from 2003 to 2007 while secretly collecting personal data, without warrants and relying on national security pretexts.[48] For its part, the FBI minimized violations as mere technical mistakes. The key problem, however, is that the Patriot Act made it easier for the FBI to gather personal information in the absence of warrants—so long as investigations could be linked to nebulous concerns of "national security." This could only have a chilling effect on political dissent, above all regarding foreign policy.

The NSS has expanded in its scope over time, under both Democrats and Republicans, whatever the variations in political rhetoric. And little has changed during the Obama presidency: military interventions in Afghanistan and Iraq continue apace, the U.S. global presence is bolstered, threats to Iran escalate. Pentagon spending exceeds one trillion dollars, and intelligence agencies can look forward to continued business-as-usual. In his book *The Obama Syndrome*, Tariq Ali writes, "From Palestine through Iraq to Iran, Obama has acted as just another steward of the American empire, pursuing the same aims as his predecessors, with the same means but with a more emollient rhetoric."[49] When it comes to the NSS, the overarching postwar motif is *continuity*. As an indispensable tool of the warfare state and imperial power, the security-state narrows the public sphere and subverts what remains of democratic politics, eviscerating reform efforts at the same time.

The Media-Propaganda Network

All developed capitalist societies, none more so than the United States, legitimate their class and power structures through ideological hegemony—a broad framework of consensual beliefs, traditions, myths, and lifestyles. As Antonio Gramsci theorized, the more industrialized, socially complex, highly educated the society—and the more established its liberal-democratic institutions—the greater the reliance on consensual rather than coercive forms of rule. Preponderant dependency on military and police force reveals nothing so much as political fragility, a shaky process of legitimation. As modern society evolves, law enforcement, the military, and bureaucratic structures generally yield to such sources of popular consciousness as education, culture, the media, and civic traditions. Writing between World War I and the mid-1930s, Gramsci could not have foreseen the full impact of newer, more sophisticated forms of hegemony such as the popular media, which exercises enormous ideological and cultural influence in the capitalist West, reinforced in recent decades by the informational revolution. American society would be at the epicenter of this shift.

By the early twenty-first century concentrated media power in the United States had become the most far-reaching of all hegemonic forms. Oligopolistic ownership, deregulation, the growth of advertising and commercialism, and a rightward ideological shift had ensured that trajectory, as expanded business control over film, TV, the Internet, radio, and print journalism led to an erosion of political debate, popular governance, and citizenship on questions of corporate power and the warfare state. Media giants (Viacom, Disney, News Corp, TimeWarner, etc.) possess formidable economic and technological resources, augmented by deregulation set in motion by the 1996 Telecommunications Act. Robert McChesney writes that "The corporate media cement a system whereby the wealthy and powerful few make the most important decisions with virtually no informed public participation."[50] Media culture shapes how major issues are framed, valorized, and contextualized, setting the permissible range of views and who can express those views. While the media often *appears* open and plural, a signpost of the "free press," those managing the communications system have enough resources to control the flow and content of what gets transmitted. The corporate media has evolved into a thriving propaganda system reaching every corner of society with technological expertise, material resources, and ideological content. While this bears little resemblance to familiar state-directed propaganda that thrives under dictatorships, it could be ultimately far more disarming and efficient. Enough has been written about the oligopolistic character of American media as to render further discussion here superfluous: the work of such writers as McChesney and Ben Bagdikian has shown how concentrated ownership and control have tightened in recent decades, enhanced by globalization, a wave of mergers and conglomerations, and the impact of the Telecommunications Act.[51] Five giant media empires dominate the public sphere, as profit-driven enterprises usher in an era in which banking, production, government, culture, education, and even the military converge.

Critics have rightly called attention to the dramatic expansion of media power as both a source of ideological hegemony and a threat to popular governance, yet it is media *content* that more decisively influences general beliefs and attitudes: a rightward evolution of media culture reflected in TV news, cable offerings, Hollywood movies, talk radio, and op-ed pages of newspapers has reconfigured the public landscape. Think tanks, foundations, lobbies, Political Action Committees (PACs), and outright corporate influence have fueled this shift, defined by stridently pro-business and antilabor views matched by support for an aggressive foreign policy, high levels of Pentagon spending, and "family values." In *The Republican Noise Machine*, David Brock describes a protracted, well-funded campaign waged by dozens of conservative groups to transform the media terrain through "savage partisanship," the breakthrough of FOX television being one result of this strategy.[52] Political views considered extreme in the 1960s and 1970s had become mainstream by the time of the second Bush presidency, more so with the Tea Party ascendancy. The neocons came to influence the outlook of both parties, reflecting a strong continuity from Bush to Obama. As for News Corp, Rupert Murdoch's right-wing media empire took initial leadership of the global-warming denial campaign, the FOX network routinely dismissing climate change as a conspiracy by liberal scientists and big government. Glenn Beck informed viewers that the planet had experienced no global warming in recent years, while Sean Hannity announced that climate change was nothing but the wild fantasy of "global-warming hysterics." Murdoch's *Wall Street Journal*, meanwhile, missed no opportunity to diminish climate change discourse as essentially apocalyptic scare tactics.

Talk radio emerged as probably the most reliable, and extreme, forum of right-wing demagoguery, with its nearly monolithic barrage of white-male attacks on the usual demons: liberals, feminists, environmentalists, peace activists, Muslims. Even the mildest critics of corporate or military power are denounced as Communists, terrorist sympathizers, Godless, traitors. The radio outlets, half of them owned by fiercely conservative Clear Channel Corp, helped energize an upsurge of right-wing populism, including the Tea Party protests that surfaced in 2009. Nearly 60 percent of all Americans were reported to be regular listeners to such talk shows. Talk radio has the special capacity to monitor inputs in order to keep disagreeable opinion off the airwaves; "debates" extend no further than differences among conservatives, above all on corporate, foreign, and military policy. Some issues (feminism and animal rights) are simply trivialized beyond recognition. Still others (U.S. military interventions) command nonstop attention but go no further than patriotic cheerleading mixed with bashing of critics. There is also the spectacle of tabloid journalism that markets celebrity scandals, true-crime sagas, popular gossip, and similar fare. Media-sponsored candidate debates, usually pitting liberals against conservatives, enter into the picture—but here too questions and answers are framed within safe ideological parameters.

It is worth noting what subject matter audiences for TV news, talk shows, and kindred venues are likely to miss—or find totally obscured. According to the volume *Censored 2010*, dozens of examples include crucial developments related to the World Bank, the IMF, the Federal Reserve, Wall Street, the global arms trade, the war on drugs, Pentagon spending, the costs of war, the cancer epidemic, and the problems of nuclear power.[53] As might be expected, probing coverage of the security-state, including activities of the NSA and the CIA, is virtually taboo. In the case of global warming, a few corporate-sponsored deniers merit equal time with scientific experts on TV news, talk shows, documentaries, or political debates. The world scientific consensus might be considered exempt from such "controversy," yet American media adheres to the fiction that each side deserves equal attention. Phony equivalence of this sort reveals how business interests can nullify challenges to their agendas, especially on matters related to the environment. Regarding foreign policy, Norman Solomon shows in *War Made Easy* how the media dutifully serves as a propaganda medium behind U.S. foreign adventures, its "coverage" defined by patriotic celebrations, its images focused on military exploits, its "news" derived from official sources and military "experts," its reporters attached to U.S. armed forces units, its journalists rarely probing White House or Pentagon claims in defense of war.[54] The press sees its work as auxiliary to military operations, a blatant but easily rationalized exception to norms of journalistic objectivity. Viewers of TV news and readers of American newspapers are treated to the same narratives: the U.S. as noble superpower reluctantly combating foreign evil in defense of human rights, democracy, and the struggle against tyranny.

While the media is widely thought to be a conduit of information and entertainment, it also functions as a sales venue for corporate goods and services. Sut Jhally argues this point, observing that "twentieth-century advertising is the most powerful and sustained system of propaganda in human history and its cumulative cultural effects, unless quickly checked, will be responsible for destroying the world as we know it."[55] In advertising discourse, Jhally notes, societal and ecological values are subordinated to business priorities of endless consumption, growth, and profits.[56] Investing hundreds of billions of dollars yearly into the constant movement of products, advertising transcends a strictly instrumental sales function as it takes on a more distinctly *ideological* role. A medium created to sell commodities, advertising—dominated in the United States by eight major firms— has created a new paradigm of media culture and, ultimately, political discourse, geared primarily to consumerism. McChesney writes, "As a driving force in our media system, advertising has brought commercial values into our journalism and culture in a manner unforeseeable in classical democratic theory and incompatible with traditional notions of a free press."[57] American electoral campaigns, for their part, increasingly revolve around TV- and media-centered frames of communication—that is, ambitious sales and marketing efforts geared to superficial, fleeting images. The media is yet another haven for corporate lobbies, largely unnoticed: dozens of business representatives, public relations agents, corporate executives, and business strategists appear regularly on TV and radio, often without being identified as such.[58] They manage public images of corporate behavior, promote goods and services, and help legitimate elite interests. Many "guests" on these outlets, cast as government, military, and business experts, are regulars: for example, health-care spokespeople critical of the 2009 Obama reforms frequently turned out to be lobbyists or consultants for Big Pharma and insurance firms.

The role of media culture in shaping popular consciousness raises questions about what exactly constitutes "public opinion" in American society—and, by extension, its relevance to the governing process. If public views, attitudes, and beliefs are so powerfully influenced by media outlets then citizen opinion would seem to be extensively managed. Wolin refers to such a phenomenon through his concept of "managed democracy," but whether this system can be regarded in any meaningful sense as democratic is rather questionable. Wolin argues that elites strive to legitimate a power structure that views democracy as a troublesome obstacle to corporate agendas.[59] "Public opinion," framed thusly, has little autonomous status, mirroring instead an ensemble of views and beliefs advanced by those who own and control the media system.[60] Domhoff refers to an "opinion-shaping network" comprising lobbies, think tanks, public relations firms, and the media—all sparing no resources to mold public views around "free markets," consumerism, big military, and patriotism, views indeed generally shared by the American public. Opinion conflicting with such consensus is either excluded from the media or marginalized as deviant. Any understanding of "public opinion" must therefore take into account the reality of an amorphous mass electorate coexisting with a system of tightly organized corporate interests and their propaganda outlets. With the media terrain so economically and politically narrowed, the ideological leverage of consumer, labor, and community groups is easily negated.

The rightward turn of American politics would surely have been impossible without the great success of the FOX network. The Tea Party insurgency and the Republican takeover of the House in the 2010 midterm elections owe much to this effective propaganda machine, reaching millions of viewers with ideologically charged messages said to be fair and objective. When Murdoch opened the network in 1996 his singular goal was to reshape the media terrain to win mass audiences behind an audaciously pro-corporate, pro-military outlook infused with "family values." With a stable of archconservative firebrands, FOX president Roger Ailes set out to transform the contours of political debate—a goal accomplished within a decade. Commentators ritually embark on tirades against liberalism (referred to as the "left"), multiculturalism, gay marriage, Hugo Chavez, immigration, Muslims, environmentalists, and, after 2008, President Obama, regularly demonized as a "Marxist" or "Communist." The pseudo-libertarian Tea Party gets round-the-clock coverage, replete with special features, interviews, and commentary by carefully groomed icons like Sarah Palin. FOX has come to occupy the apex of a right-wing propaganda apparatus, a crucial part of what McChesney and John Nichols call the "money-and-media election complex."[61]

As media giants expand their scope, further restricting public access and content diversity, mergers within the telecommunications industry—fueled by deregulation—have enlarged the power of behemoths like AT&T and Verizon. By 2008 the top four Internet providers (Comcast, TimeWarner, Verizon, AT&T) controlled nearly 60 percent of all online subscriptions. In 2010, Comcast maneuvered to purchase NBC in one of the largest buyouts in history. NBC, of course, was already a major network with interests in movie studios, TV, radio, and cable as well as manufacturing. Meanwhile, online video services were controlled by such giants as Google, Microsoft, and Yahoo. The aim of mergers and consolidations, like corporate power in general, is to maximize "private" leverage over increasingly globalized communications markets—that is, over media content, distribution, and audience. As for the Internet, the main corporate objective is to restrict competition over the flow of digital information, only partially achieved (as of 2011). Meanwhile, journalism in the United States continues to deteriorate: if the news enterprise was traditionally regarded as a public good, in recent decades that "good" has been delivered to business interests geared to profit maximization. As Nichols and McChesney write, " . . . if business-as-usual prevails, we face a future where there will be relatively few paid journalists working in competing newsrooms with editors, fact-checkers, travel budgets, and institutional support. Vast areas of public life and government activity will take place in the dark—as is already the case.... Independent and insightful coverage of the basic workings of local, state, and federal government, and of our many interventions and occupations abroad, is disappearing as rapidly as the rainforests. The political implications are dire."[62] The familiar "business model" of media has indeed failed miserably, from the standpoint of both public interest and journalism. Newspapers, for their part, now survive only through mergers, where press functions are joined together under broader ownership. The business side is compromised by drastically falling revenues as readers gravitate to the Internet, reinforcing pressures toward consolidation. Even where papers can retain their journalistic identities and own editorial boards, the trend toward combining operations (news, advertising, sales, distribution) is difficult to resist. A wide stratum of hedge fund operators hovers over the media scene, always ready to drive the trend toward concentrated power, further harming media integrity and content.

A mass media owned and managed by megacorporations, wedded to commercial venues, will predictably be in the vanguard of global-warming denial. The period leading to the 2010 Cancun gathering on climate change saw meager coverage in the United States: news stories were brief, casual, and relegated to the back pages. TV coverage was hard to detect at FOX, CNN, NBC, and other major networks. Even here, the media turned almost exclusively to *business* implications, giving short shrift to the magnitude of the crisis. Stories in the *New York Times* and the *Los Angeles Times*, two of the few American papers sending reporters to Cancun, described the meetings as boring and meaningless, a statement about U.S. priorities as much as the level of interest generated by the conference. Media disinterest in the face of scientific consensus around global warming coincides with the larger political impasse on environmental issues. Scarcely visible within TV news programs, talk shows, radio, and print outlets, the threat of climate change ranks far behind a long list of concerns, real or contrived, that occupy media culture. More revealing, the issue of global warming was never taken up during the 2010 midterm elections, either in debates or on the campaign trail. Lost in the discourse around taxes, public spending, and deficits was a focus on an ecological crisis that is, in the final analysis, dialectically interwoven with the economic predicament.

Nichols and McChesney note that, while the Internet and particularly its social-networking dimensions are trumpeted as great democratizing advances, the actuality is something altogether different insofar as online traffic sites resonate with corporate interests. Web sites, nominally open and broadly accessible, are in fact managed by the usual suspects identified earlier.. While the Internet does favor diversity, an optimistic new phase of "digital democracy" has been overstated, even as the role of social networking in the 2011 upheavals across North Africa and the Middle East cannot be denied. The new technology clearly serves as an indispensable tool of social communication and political action, but it is far too soon to make firm conclusions about its long-term impact for popular movements and political revolutions. For one thing, the ruling elites have access (surely *greater* access) to the same technology, permitting enhanced information gathering, networking, communications, and surveillance at the highest levels of power. Further, while technology provides advantages in the realm of communications, its capacity to maintain community building and other necessary elements of durable politics (organization, ideology, strategy, etc.) remains unsettled. It could be that sustained political action thrives better on a more low-technology, human-centered approach like that involved in old-fashioned community organizing. In any event, there is little evidence to date in support of a historically novel (and politically successful) "twitter revolution."[63] The vast money-and-media complex (partially including the Internet) has in fact turned out to be far more problematic for both media culture and political discourse, not only in the United States.

Electoral Delusions

The United States has a long history of relatively free and open elections founded on a Constitution and institutional pluralism ritually celebrated by politicians, the media, the educational system, and the general population. This tradition, however, can be deceiving. When electoral and legislative activity is framed against the backdrop of class and power relations, the governing mechanisms—however democratic in theory—can easily be controlled for the benefit of elites with resources (wealth, organization, skills, social networks, etc.) to do so. Electoral politics in itself does not fundamentally alter the structure or conduct of elite power, especially where forms of domination are well established. Nor does electoralism necessarily provide a vehicle of social reforms, though it clearly has that potential. Behind the mix of parties, elections, and parliamentary activity lies the corrosive influence of corporate and financial power, well-funded lobbies, the warfare state, and bureaucratic structures across the public terrain.

The power elite has accrued new levels of institutional leverage and ideological hegemony throughout the postwar years, as Mills had foreseen in the 1950s. Drawing on the Marxist tradition and the work of elite theorists like Gaetano Mosca and Robert Michels, Mills depicted modern liberal democracy as an intricate and seductive device for legitimating the status quo. And Mills was essentially correct: across U.S. history ruling elites have been the most politically engaged sector of a dominant stratum based in manufacturing, banking, agriculture, other large-scale businesses, and, more recently, the military and intelligence. Today these elites are deeply entrenched not only in the system of production but in banking, insurance, pharmaceuticals, energy, the media, universities, and the Pentagon. Whether in federal, state, or local politics, Congress or the presidency, Republicans or Democrats, wealth has always shaped political life in America, betraying familiar pretenses of equality, democracy, and pluralism. One useful maxim is that the more power amassed by corporations, the more likely popular access, accountability, and citizen participation will be subverted. A supposed neutral arbiter of multiple and freely competing interests, in fact the state-capitalist apparatus constitutes a mighty façade behind which democratic ideals are routinely compromised within everyday politics.

The American two-party system is not so tightly integrated as to nullify differences on social issues or conflict over questions such as taxation, corporate regulations, health-care reform, welfare spending, and foreign trade. It is, however, integrated enough to block alternatives to established corporate, financial, and military agendas. To win elections, politicians of both parties compete for the same sources of campaign financing and are beholden to the same interests, despite variations in campaign rhetoric and policy choices. Single-member, winner-take-all electoral districts obstruct third-party challenges while Republicans and Democrats, always in search of majorities, gravitate toward the "center" and away from "extreme" positions. Political success depends on a candidate's ability to raise millions (even tens of millions) of dollars in "private" funding. The pressures of corporate financing and media advertising dictate that campaigns follow three imperatives—focus on personal characteristics, embrace of ideological platitudes (peace, growth, family values, etc.), and support for corporate and military priorities. Candidates who deviate from this formula are sure to be defeated. Congress itself is often paralyzed by legislative impasse, with power dispersed among myriad committees, a bicameral Congress skewed toward rural and small-state interests, and a parliamentary maze dominated by moneyed interests and their lobbies. Debates over gay rights, abortion, gun control, and taxes usually consume the bulk of media and political attention. Differences over environmental legislation can be heated, but political agreements limiting reforms to corporate-friendly parameters of growth and profits typically prevail.

American politicians invariably come from the top 10 to 15 percent on the income ladder, with legal and professional backgrounds prevalent and white males (75 percent of the Senate) still a majority. Few enter the corridors of power from poor or working-class backgrounds.[64] Business funding, as noted, supports thousands of lobbies in the Beltway and state capitols. From 1998 to 2009 the leading federal spenders were the Chamber of Commerce ($527 million), the American Medical Association ($212 million), and General Electric ($191 million), followed by PhRMA ($161 million), AT&T ($151 million), Northrop Grumman ($144 million), and ExxonMobil ($131 million). In 2009 alone Washington lobbies spent more than $2.5 billion to shape legislative outcomes.[65] Democrats and Republicans were roughly equal recipients of corporate largesse. The 2008 presidential campaign broke all spending records with the "populist" Obama raising $730 million, more than double Republican John McCain's $333 million. Obama's fund-raising coup was made possible by such firms as Goldman Sachs ($995,000), Microsoft ($834,000), Google ($803,000), Citigroup ($701,000), JPMorgan Chase ($695,000), TimeWarner ($590,000), IBM ($518,000), and Morgan Stanley ($514,000).[66] Thus Obama's grassroots efforts to raise money through the Internet, while impressive, were dwarfed by the larger plutocratic tide. In fact the three leading Democrats after 2008 (Obama, House Speaker Nancy Pelosi, Senator majority leader Harry Reid) owed their political fortunes to heavy corporate backing. Instead of a populist assault on the centers of financial power responsible for the economic meltdown, therefore, Obama and the Democrats ended up content with business-as-usual.

The 2010 national midterm elections even more vividly highlighted the plutocratic cast of American politics. Republicans won a series of heated congressional battles, wresting control from House Democrats—inspired by Tea Party promises to fight the "establishment" and roll back "big government." While Tea Party – backed candidates lost the bulk of contested Senate seats, Republican takeover of the House (and many state legislatures) was driven by a contrived image of "outsiders" fighting the corrupt Beltway culture. Senator Jim DeMint (R-SC) claimed that "Tea Party Republicans were elected to go to Washington and save the country, not be coopted by the club."[67] This sentiment echoed the right-wing consensus, which viewed Obama as delivering the country to "socialism" or some nightmarish equivalent. Republicans made gains on a platform of small government, lower taxes, deficit reduction, and loosened business regulations. One problem with this "insurgency," however, was its fraudulent character: hardly a triumph of antiestablishment "outsiders," Republican advances were a victory for big business and Wall Street, the beneficiaries of deregulation, tax cuts for the wealthy, and increased military spending. The most expensive midterm election in U.S. history, its outcome signaled a more profound oligarchic turn, a corporate-backed assault on social programs and public infrastructure masked as populism. Tea Party candidates relied on an infusion of big-business funding through the Chamber of Commerce, Karl Rove's Crossroads GPS, and Americans for Prosperity as well as the multibillionaire Koch brothers, backers of the Tea Party in its crusade to expand the elite power of big business, finance, the military, and the wealthy. The notorious 2009 Supreme Court decision, *Citizens United vs. FEC*, allowing corporations to spend unlimited amounts on electoral campaigns, only further skewed the electoral process.

Matching the Tea Party love affair with big corporations and banks is a worship of the same big government it so hypocritically attacks. Archconservatives scream that government is too big, has too much control over people's lives, stifles the "free market," and tramples individual freedom, yet lend uncritical support to a bloated state machinery for their own tax-and-spend schemes: the war economy, the security-state, federal bailouts, a massive intelligence apparatus, a prison complex, and the disastrous war on drugs. About these schemes, which drain trillions of dollars from the public treasury and run up the public deficit, the great libertarians and "populists" remain silent. No Tea Party leaders have seriously discussed reducing a military budget exceeding that of all other nations combined. It turns out that "tax-and-spend" is an exemplary practice when it serves elite interests, but scandalous when it funds public goods like health care and the environment. Such conservatives are happy to spend profligately on an already bloated war economy and security-state, not to mention armed interventions abroad, comfortable with a regimen of global ventures and righteous warfare.

The Tea Party ascendancy within the Republican orbit signals the most savage attack on the public sector since the New Deal reforms were enacted in the 1930s—an attack supported, in varying degrees, by leaders in *both* parties. Couched as a discourse around the "budget," "deficits," and "big government," this revitalized libertarianism embraces corporations and the wealthy as heroic producers of prosperity, jobs, and freedom while taking aim at workers, unions, consumers, and government as a locus of slackers and "parasites," a drain on growth and affluence. Fueled by growing interest in the work of Ayn Rand and peddled by right-wing think tanks, this ersatz libertarianism depicts "liberal government" as tyrannical, evil, and corrupt, its programs representing a theft of public revenues. But such a libertarianism, funded by corporations and billionaires, represents a barely concealed class warfare of the rich against labor, the poor, and even the middle classes. Rep. Paul Ryan of Illinois, a great champion of Tea Party ideology and architect of severe public cutbacks, argues that Rand was the greatest philosopher who ever lived as he presses for "small government" and increased "personal responsibility." Like the Tea Party in general, however, Ryan exudes deceit and hypocrisy. In recent years he, like the vast majority of his Republican colleagues, voted for every massive government spending program: the Pentagon, intelligence, Homeland Security, wars in Iraq and Afghanistan, law enforcement, federal bailouts. Moreover, he did this while supporting Bush's tax breaks for the rich. It was only after Obama's entry into the White House that Ryan suddenly discovered the terrible menace of "fiscal deficits" and out-of-control public spending. It follows that the real Tea Party agenda, far from attacking "big government," is to shelve any possible reform legislation in the name of "debt reduction" and "fiscal restraints."

If Tea Party Republicans and their allies succeed in gutting the public sector, federal and state protections against the predation of corporate and Wall Street power will sadly become a faded memory. The crusade is financed by such super-wealthy "libertarians" as the Koch brothers, whose fortunes (at least $40 billion combined) are derived from the energy and chemical industries. They have spent more than $200 million funding conservative think tanks since the late 1990s, another $35 million to defeat climate change legislation, and yet another tens of millions for the candidacy of hundreds of politicians since 2005. Behind efforts to deregulate corporate interests and eviscerate the public sector, their mantra is "small government," "free markets," and "personal responsibility," ostensibly in the service of prosperity, jobs, and national security. The Koch brothers have poured tens of millions of dollars into think tanks like the Cato Institute, well known for its "research" demonstrating that global warming is a "hoax," and such advocacy groups as Freedom Works and Americans for Prosperity. The brothers have also funded ALEC, working to set election barriers to students, immigrants, ex-felons, and others likely to vote Democratic—part of a "war on voting" that could disenfranchise millions of potential voters in state and national elections. Inspired by Koch-backed activism, no less than 38 states (in 2011) were considering some form of legislation (such as identification cards) to discourage or block citizens from voting.[68]

The historic merger of corporate and governmental power endemic to advanced state capitalism has come to define the "American model," in which popular sovereignty, citizenship, and social governance essential to democracy are reduced to abstract ideals little related to the exercise of power. Neither major party even *desires* an alternative to oligarchic rule, hardly surprising given that both have evolved into stable fixtures of corporatized, institutionalized power. American liberalism, earlier understood as a Keynesian alternative to rampant class exploitation and capitalist breakdown, has eroded in the face of the well-orchestrated conservative crusade. Chris Hedges argues in *The Death of the Liberal Class* that the historical pillars of liberal politics (unions, academia, Democratic Party, community organizations) decline as elites move to assault the last vestiges of reform and regulation.[69] Even tepid attempts to soften the harsh edges of an aggressive, freewheeling capitalism today appear as intolerable threats to elite wealth and power—witness the hysterical opposition to Obama's weak health-care and financial reforms. Hedges writes, "In killing off the liberal class, the corporate state, by shutting down reform mechanisms, has created a closed system defined by polarization, gridlock, and political theater. It has removed the veneer of virtue and goodness provided by the liberal class."[70] Liberal erosion leaves a political void filled by financial speculators, war profiteers, media hucksters, and Tea Party frauds as party competition further degenerates into a commodified spectacle.

As the political culture moves ever rightward—meaning the power elite is freed to better carry out its exploitation, fraud, criminality, and propaganda—efforts to reverse the global ecological crisis will meet burdensome obstacles, reflected in the House-inflicted 2011 political impasse over the debt ceiling. The American electoral system is designed to block meaningful change, so that Obama's failure to uphold a liberal agenda, including promised environmental reforms, comes as little surprise; corporate and military interests exercise such hegemony that no White House occupant could hope to chart an independent course. Tariq Ali, in *The Obama Syndrome*, writes that "A modern American president—Republican or Democrat—operates as a messenger-servant of the country's corporations, defending them against their critics and ensuring that no obstacle be placed in their way."[71] Obama's decision to staff his administration with Wall Street insiders and Pentagon hawks should therefore have been easy to anticipate. Ali points out that, "In reality, Barack Obama is a skillful and gifted machine politician who rapidly rose to the top. Once that is understood there is little about him that should surprise anyone: to talk of betrayal is foolish, for nothing has been betrayed but one's illusions."[72]

Although environmentalists worked hard to get Obama elected in 2008, optimistic that bold measures might be on the agenda, by late 2011 all such promises and hopes were reduced to ashes. In fact the new administration, preoccupied with the economic crisis and facing extreme Republican opposition, appeared to lose all interest in environmental legislation. Republican agendas quickly surfaced, especially following the 2010 midterm election: renewed attention to nuclear energy, development of "clean" coal, more natural gas, endorsement of the aforementioned Canada-U.S. pipeline, renewed offshore oil drilling, opening of more public lands to coal mining—basically a continuation of the Bush agenda. As for global warming—virtually nothing, Obama doing little to assist passage of the 2010 climate change bill that ended up stillborn. In public speeches Obama rarely got around to even *mentioning* the crisis, with public remedies off the radar. This was the same president, beholden to corporate lobbies and Wall Street funding, who failed to address the gross political, legal, and ethical violations of the financial sector that occurred under Bush. Such continuity stems not so much from Democratic "retreat" or capitulation to ruling interests as from a long-standing dedication to business-as-usual within the corporate-state.

As the 2012 presidential election rolled into view, oligarchical tendencies gained new headway. The role of the super-rich in American politics emerged with greater clarity as "super-PACs" went into full swing: Restore Our Future, for example, backed Mitt Romney's Republican candidacy with some $30 million during 2011, enhanced by million dollar donations from New York-based hedge fund executives—Paul Singer, Robert Mercer, and Julian Robertson.

A committee backing Newt Gingrich kept the former House Speaker's Republican hopes alive with massive contributions from gambling magnate Sheldon Adelson. The fiercely pro-Israel Adelson and his wife gave $10 million to a Gingrich-allied super-PAC while other relatives gave a total of one million dollars to the Winning Our Future fund. Dallas billionaire Harold Simmons was competing with Adelson as the most generous donor of the 2011–12 electoral season, having given seven billion dollars to American Crossroads, the super-PAC founded by Karl Rove. In fact American Crossroads disclosed that it received more than $18 billion, with five billion dollars coming from Simmons personally and two billion more from his holding company, Contran Corp. Overall, Republicans are recipients of more than double the contributions amassed by the Democrats.[73]

While integrated corporate-state power narrows politics and deflates citizenship, traditional ideologies (liberalism and conservatism) grow stale and irrelevant, their main function being to legitimate elite interests. And while giant corporations shape the flow of capital, natural resources, commodities, information, and technology, their reach evades the orbit of parties and elections.[74] As oligarchic trends solidify, we find a political culture of phantom discourses, rife with illusions and fantasies, often favoring popular retreat and cynicism—signs of a social order in severe decline. As the United States contributes its lethal share to global warming, using 25 percent of world energy resources, Washington politicians continue to stonewall measures needed to confront the ecological crisis. The United States stridently opposes even slight alterations of business-as-usual, whether at Rio or Kyoto, Copenhagen, and Cancun. Limits to growth and profits are resisted across the public landscape, from the Congress to the White House, state legislatures, the Supreme Court, the media, and even mainstream environmental groups. Writing that "Our environment is being dramatically transformed in ways that will soon make it difficult for the human species to survive," Hedges adds, "We stand on the verge of one of the bleakest periods in human history, when the bright lights of civilization will blink out and we will descend for decades, if not centuries, into barbarity."[75] Unfortunately, life-affirming alternatives to an ecologically ruinous corporate state will have to come from outside, and against, the deeply embedded networks of power, wealth, and Empire.

Chapter 4

Liberal Delusions

The previous chapter dealt with imposing difficulties of winning genuine reforms within an institutionalized corporate-state that diminishes the realm of politics at a time when American society moves further along the path of destructive and costly modes of production, consumption, and lifestyles. In this setting appeals to "realism" and "pragmatism"—historically resonant in American public life—essentially mean capitulation to the dominant interests: recent "solutions" to the global crisis offered by liberal environmentalism fit this very pattern. An expression of early capitalist development, classical liberalism promised a new era of equality, democracy, and prosperity inspired by Enlightenment values, but with the passage of time such expectations were at best partially and unevenly realized. By the late twentieth century the liberal tradition had become associated with a lengthy period of sustained economic growth, yet modern corporate liberalism would be a signpost of sharp inequalities, truncated democracy, and affluence for a shrinking minority, while pushing society toward environmental ruin. In the United States, moreover, the liberal-capitalist predicament was heightened by the expansion of a war economy and a security-state requiring burdensome costs and resources. Yet this system, whatever its problems and obstacles, managed to achieve substantial popular legitimacy, with political opposition normally confined to fixed parameters. The two-party system, electoral activity, interest groups, and the legislative process were extensively colonized by corporate interests, a supposed vehicle of free markets, economic growth, democratic governance, and global peace. The urgency of ecological challenges, however, was not lost within the political establishment, its rumblings giving rise to a number of liberal schemes, the most ambitious including the work of Al Gore, Thomas Friedman, and Lester Brown.

THREE FACES OF LIBERALISM

Former vice president Al Gore has long believed that the environmental crisis poses the defining challenge for humanity, a central motif of his 1995 best-selling book, *Earth in the Balance*. Since leaving office in 2001, Gore has devoted the bulk of his time and energy to the threat of global warming—the topic of his documentary *An Inconvenient Truth* (2006), which later brought both an Oscar and a Nobel Prize. Gore's views are perhaps most systematically elaborated, backed by the generous use of charts and graphs, in his 2009 volume, *Our Choice*, dedicated to raising public consciousness about the perils of climate change.[1] Gore's body of work furnishes one of the most comprehensive overviews yet of the global challenge—empirically grounded and driven by intellectual passion and sense of political urgency. His main objective is to present scientifically valid information, consistent with the aforementioned IPCC reports, countering the massive distortions and denials of corporations, their bought politicians, their lobbies, and the media. Americans are exposed to pervasive narratives of doubt and skepticism amidst the "balanced" treatment of views on questions long ago resolved in world scientific consensus. For Gore, major obstacles to reversing the crisis reside first in *ideological* discourses fashionable in the media, lobbies, and think tanks created to serve elite agendas. Lacking reliable information and analysis, Americans have grown calloused to a social order immersed in hyper-consumption, waste, and environmental devastation, seduced by the myth that prosperity and happiness depend on endless material growth, made possible by "free markets," a system that indeed has produced close to a *tripling* of domestic output since the 1950s.

The subtitle of Gore's book is, "A Plan to Solve the Climate Crisis." His idea is to chart a workable strategy of environmental reform, anchored to the notion that the United States remains at the epicenter of the global crisis and therefore must be a key site of any future green transformation. Despite a mood of emergency conveyed in the text, however, the proposed strategy—seemingly ambitious to the core—departs little from standard liberal politics. In the end, Gore's understanding of social change is rather soft and tentative, well short of the anticorporate regimen needed for a strong ecological politics. He looks to a creative mixture of solutions: education, green technology, new market initiatives based on environment-friendly investment, lifestyle transformations. The spread of knowledge and awareness, of course, is indispensable to fighting corporate propaganda and debunking phony science. People must be convinced, and soon, that nothing short of human survival is at stake—a difficult task given the extent to which corporate narratives saturate media culture. As might be expected, Gore exudes unwavering faith in the power of technology to drive alternative sources of energy and, ultimately, developmental sustainability. The same computers and supercomputers that integrate and transmit vast flows of information while promising revitalized democracy can assist pathbreaking green-technology designs rooted in solar, thermal, wind, and other less destructive sources of power.[2] Education and technology thus merge as crucial instruments for environmental reconstruction.[3]

As with *Earth in the Balance,* Gore's ecological outlook leans heavily on market "solutions," despite harsh words for loosened corporate regulations and "short termism," or obsession with next-quarter profits. Throughout *Our Choice* a jaundiced view of corporate power is evident, yet, consistent with his establishment credentials, a liberal reformism holds sway, spurred by market efficiency and material incentives. The optimum strategy, he argues, is to fuse public and private initiatives within a developmental model accounting for the costs of carbon extraction, processing, use, and pollution. Along with green technologies to reduce emissions, Gore favors a carbon tax to supplement cap-and-trade schemes, introduced at Kyoto, restricting carbon pollution through market-based trading arrangements. The idea is to identify, measure, and assess greenhouse costs while utilizing existing mechanisms like the Clean Air Act.[4] Despite misgivings about capitalism and its sharpened exploitation of the natural habitat, therefore, Gore is unambiguous: "We must develop a sustainable capitalism."[5] Even as corporations and their lobbies struggle tenaciously to deflect or block reforms, their *modus operandi* can be turned around to advance ecological rationality. Thus, "Our market economy can help us solve the climate crisis problem if we send it the right signals."[6] The "right signals" would be adequate carbon-pricing mechanisms combined with green-technological incentives. This is an epic project, to be sure, akin to a new Manhattan Project that, once implemented, would reestablish U.S. "moral authority" as world economic and technological leader.[7] Gore's perspective, here as elsewhere, resonates with an unshakeable Enlightenment optimism: corrupt and dysfunctional as the system might be, it can be steered in the right direction by means of creative leadership, good intentions, and improved priorities.

In his best-selling *Hot, Flat, and Crowded*, Thomas L. Friedman makes the case for a "green revolution" to correct the tendencies of a dangerously unstable and crisis-ridden planet. A *New York Times* writer, Friedman proposes a survival modality built around the same three instruments embraced by Gore—alternative technology, revitalized markets, and lifestyle changes. The case for such a dramatic turnaround, no longer simply an option, is to be made on moral, political, and economic grounds, involving a program of national renewal along lines of the American World War II mobilization for battlefield victory.[8] One impediment to this shift is that mainstream environmental movements have yet to transcend the feel-good ethos of Earth Days, rock concerts, films, and celebrity testimonials.

Friedman's view of the global crisis is relentlessly grim—a narrative of decaying infrastructure, vanishing natural resources, fossil-fuel addiction, broken political systems, and populations anxiously facing uncertain times ahead. Change must occur, and soon, requiring not only ambitious alternatives to existing patterns of energy use but more urgent public commitment to planning for the future, meaning overthrow of market fundamentalism and more vigorous government intervention. Friedman argues that clean power driven by green technology will underpin the next major world economic stimulus, furnishing abundant, cheap, reliable, and sustainable fuel sources crucial to saving the planet.[9] Given the high levels of American industrialization, the lopsided U.S. contribution to global warming, and U.S. world economic and military supremacy, Washington decision making will be crucial in pushing the world along a green path—or blocking that path. Friedman compares present-day ecological challenges to the war against the Nazis and to the "Communist threat" during the Cold War. With smarter, cleaner, and more efficient technology to power electricity and transportation, it will be possible—assuming strong market incentives—to sustain perpetual economic growth, with no reduction in living standards or domestic production necessary.[10] Technological innovation is destined to reshape every sphere of human activity: manufacturing, agriculture, communications, education, transportation, health care, even the military. Rejecting the limits-to-growth ideology, Friedman anticipates future economic development without limits, infused with a green regimen as solar, wind, geothermal, and fuel-cell energy sources gradually replace outmoded fossil fuels. Such an alternative, moreover, can be relatively painless and inexpensive, all parties ending up as winners. Convinced that environmental sustainability and economic growth are not only compatible but mutually reinforcing, Friedman looks to the greening of personal lifestyles, extending to work, transportation, and consumer choices. Sustainable values would be integrated into the rhythms and flows of family and individual choices. People will reduce their carbon imprint by driving less, buying more fuel-efficient vehicles, using alternative energy sources, reducing electricity needs, and joining community projects like recycling, car pooling, and tree planting. Believing ecological sustainability is compatible with ideas already espoused by leading Democrats and Republicans, Friedman refuses discussion of whether and to what degree vested interests must be confronted or overturned.

The ecological "revolution" Friedman envisions is eminently possible within corporate-state parameters—presuming, of course, large-scale changes in technology, the market, and personal lifestyles. He believes that homes, like transportation and utility systems, can be rebuilt or renovated to meet post-carbon standards—made cleaner, cheaper, and more efficient through renewable power sources.[11] As homes, businesses, schools, universities, and other centers of life enter into this transitional process, new entrepreneurial venues are sure to appear, generating ever-more dynamic ecological mechanisms, Friedman adding, "...the one thing that can stimulate this much innovation in new technologies and the radical improvement of existing ones is the free market."[12] Thousands of companies, big and small, are to be assimilated into the green transformation. While government imposes taxes, provides subsidies, and enacts social policies (e.g., the carbon tax), Friedman embraces the "market" as the main vehicle of innovation and growth, as investment choices within capitalism furnish all that is indispensable so long as the proper "signals" are transmitted.[13] The central role of state power, in fact, is to encourage more competitive and thriving markets. By heavily taxing gasoline, for example, government can stimulate production of more efficient cars as well as improved modes of public transportation. The spread of green technology, Friedman argues, will position the United States as international leader in sustainable growth, the driving force behind a new (ecological) "American model." While trumpeting virtues of the "free market," however, he concedes a dynamic role for the public sector, since the market alone cannot set conditions for research, investment, pricing, and sales.[14] The current problem, for Friedman, is that U.S. energy policy is "incoherent, ad hoc, and asystematic," in contrast to that of most European governments.[15]

The subtitle of Lester Brown's important *Plan B 4.0* is "Mobilizing to Save Civilization," a book written as a comprehensive blueprint for reversing the global crisis. President of the Earth Policy Institute and a driving force of modern environmentalism, Brown leaves no topic untouched in his plea for a sustainable future. In the face of impending catastrophe, he urges a "massive mobilization to restructure the world economy," to be completed "at wartime speed" as if facing another world war.[16] Ecological revolution calls forth a total political commitment based on new priorities, large-scale planning, and collective mobilization. Brown in fact sets forth a more complex and variegated strategy than either Gore or Friedman, with four interrelated components: climate stabilization, limits to population growth, eradication of (world) poverty, and restoration of natural support systems.[17] With the tipping point near, change requires fundamentally new modes of thought and behavior, Brown concluding, "Like earlier civilizations that got into environmental trouble, we can decide to stay in business as usual and watch our modern economy decline and eventually collapse, or we can consciously move onto a new path, one that will sustain economic progress. In this situation, the failure to act is a de facto decision to stay on the decline-and-collapse path."[18]

Plan B is more far-reaching than what any nation has so far undertaken, all the more urgently needed because "climate change poses a threat to our civilization that has no precedent."[19] Brown insists that carbon emissions can be reduced a staggering 80 percent by 2020, well beyond anything proposed (much less accepted) at Rio, Kyoto, Copenhagen, or Cancun. New investment priorities hinged to radical technological innovations would fuel a revitalized economy powered increasingly by solar, wind, geothermal, and comparable green sources. Adopting strict carbon taxes, cities, businesses, and public services could undertake more sustainable modes of production and consumption. Market incentives would coincide with governmental, corporate, work, and lifestyle transformations keyed to a worldwide greening process. Electric cars would of course predominate. Clean and efficient systems of public transportation would supplant autos as the preferred and easier travel option. Fossil fuels would rapidly become obsolete. Green cities would be rationally designed, replacing the dysfunctional features of cities that are now sprawling, chaotic, violent, congested, and unsustainable. In Brown's view, the starting point of environmental rebirth is a downsizing and reconfiguring of vast megacities that today (2011) house more than three billion people. These monstrosities already require new infrastructures for transportation, housing, water, electricity, public health, and food that, in the future, could be sustained by alternative energy forms. Cities will have to be renovated to suit human needs, meaning greater localism, smaller and less dense populations, greener environs, more efficient public transportation, expanded car-free zones, and community-based agriculture. If cities are to fully coexist with surrounding ecosystems, then full-scale rethinking of the urban experience is vital, starting with more efficient utilization of resources. One intractable problem, Brown notes, is that "cities require a concentration of food, water, energy, and materials that nature cannot provide." In the end, "collecting these masses of materials and later dispersing them in the form of garbage, sewage, and pollutants in air and water is challenging city managers everywhere."[20] In a world with nearly one billion private cars, a drastic shift in transportation patterns becomes especially critical.

Extending his environmental outlook beyond that of Gore and Friedman, Brown argues that solutions to the agricultural predicament are fundamental to any viable ecological politics: today food resources are barely adequate for world populations, much less the eight or nine billion people expected to inhabit the planet by 2050. Soil, water, and other resources—for example, petroleum for fertilizers, pesticides, transportation, and processing—cannot satisfy escalating food demands, leaving aside the potential horrendous effects of climate change. Worldwide arable land is shrinking, and much of what remains is eroded or polluted. The international grain harvest has been relatively flat since 2000, in the midst of increasing human needs and with nearly 40 percent of all grains being fed to animals. Underground water aquifers are drying up at alarming rates. Global meat consumption reached nearly 300 million tons by 2010, devouring even more of the grain harvest. Not only is the meat industry intolerably costly and resource depleting, it hastens global warming as a leading source of methane emissions and carbon pollution. Brown's approach, to his great credit, envisions a drastic shift from animal-based to plant-based diets combined with more localized agriculture, such as urban farms, vital to reducing carbon emissions and resource consumption.

Brown sees the agricultural dilemma as tightly connected to economic sustainability issues, resource utilization, global warming, health problems, the political challenge, and population pressures, noting, "In a world where cropland is scarce and become more so, decisions made in ministries of transportation on whether to develop land-consuming, auto-centered transportation systems or more diversified networks, including light rail, buses, and bicycles that are much less land-intensive, will directly affect world food security."[21] Financial speculation over food commodities, which drives up prices, also enters the picture. Brown is relentless on the issue of population, even as observers across the political spectrum rise to debunk the challenge: if two billion people go hungry under *present* conditions, how can planetary resources feed eight or nine billion people in the decades ahead, especially with the onset of peak oil and worsening climate change? With some 80 million new people to feed annually, the world cannot hope to meet the challenge of food scarcity, especially given the limits of corporate, meat-centered agriculture. Yet Brown's prescription is hardly radical—halting population growth at eight billion by 2050—although even this "will require an all-out population education effort to help people everywhere understand how fast the relationship between us and our natural support systems is deteriorating."[22]

Gore, Friedman, and Brown exemplify the best of liberal environmentalism, testing the limits of what had been previously deemed eminently "practical" or "realistic." Allowing for some nuanced differences, their work shares an action orientation along four overlapping fronts: economic restructuring, technological innovation, lifestyle changes, public sector initiatives. All look to something resembling a new Manhattan Project, heavily dependent on government planning and resources, despite vagueness on political initiative and methods. All agree that time is very short, that the world has reached—or will soon reach—an ecological point of no return. Further, all believe that the United States stands at the epicenter of the crisis because of its vast wealth, resources, and primary contribution to global warming. As liberals, however, they cannot escape the ideological boundaries set by the corporate state, firmly convinced that reform measures achieved through existing institutional processes can be the foundation of an "ecological revolution." (Only Brown goes so far as to emphasize the need to confront the U.S. war-fare state.) There is no transformative strategy, no full embrace of an *alternative* (ecological) mode of development, no policies or goals adequate to subverting business-as-usual.

Market Fantasies

The central question posed by liberal environmentalism is whether global capitalism (or any national capitalism) can be reformed in ways that will allow political forces to take up the ecological challenge—that is, whether the approaches of Gore, Friedman, Brown, and other liberals can lead to developmental sustainability. The evidence to date—and it is abundant—is that the corporate-state is not deeply reformable enough given its stifling oligarchical character, its logic of profits and growth over public interests, and its formidable structural and ideological barriers to change. As capitalism and liberalism are historically linked, efforts to reshape corporate agendas within liberal parameters inevitably run up against limits imposed by capitalism. Since Gore, Friedman, and Brown rely so heavily on reconfiguring market priorities, they wind up trapped in this logical impasse: while modern capitalism, as in European social democracy, is surely open to greening and other reform initiatives, its very *modus operandi* conflicts with moves toward ecological rationality. Reform measures will not preclude world capitalism from following a trajectory of growth and profits leading to crisis and breakdown, nor will such measures empower those popular forces needed to avoid this scenario.

Despite their differences, Gore, Friedman, and Brown share an abiding faith in "the market"—a mechanism that, governed through reform-oriented purposes and incentives, can shed its dysfunctions and allow for sustainable development. Friedman claims that the market will give humanity everything needed to transmit greening signals to the economy, assuming high levels of technological renewal, noting that " . . . the one thing that can stimulate this much innovation in new technologies and the radical improvement of existing ones is the free market."[23] The market appears as such a fixed element of American life that alternatives seem unimaginable, beyond the scope of debate. Liberals frequently argue that nonmarket options will give rise to something akin to the Soviet command economy, forgetting that the "free market" exists nowhere—that capitalism today is generally statist, with its well-entrenched oligopolies, corporate globalization, an unaccountable financial system, a war economy, and extensive government involvement across the public landscape. Viewed thusly, the market has always been resolutely hostile to far-reaching changes in the modes of production and consumption. Could existing arrangements, so organically tied to the commodification of everything, suddenly and totally reverse course enough to satisfy pressing ecological imperatives? The problem is that liberalism, no less than conservatism, is trapped in a matrix of fictions, myths, and denials, not least of which is the naive belief in "free markets" and their simplistic linkage to democratic politics and public goods.

The phenomenon of "greening" has in fact turned into an ambitious corporate strategy, geared not only to less waste, more efficiency, and greater returns but to a project of legitimation: eco-friendly facilities and products are viewed as a sign of goodwill and social enlightenment, vital to a twenty-first-century advertising and growth agenda. In *Climatopolis*, Matthew Kahn argues that, while capitalism has done much to destroy natural habitats, if properly restructured ("greened") it can revive and save the environment while it restores its own status and efficacy.[24] Some of the largest corporations have taken the lead in market-based alternative technology, crucial to marketing and sales campaigns for such businesses as IBM, ExxonMobil, Dow Chemical, and Wal-Mart. Today literally thousands of products sold at retail outlets are routinely touted as "natural," "green," and "organic," all marketed through ecological packaging. Greening of this sort can boost sales, reduce energy costs through clean technology, and set forth an image of planet-saving business practices that, in all other respects, fit the corporate (growth-oriented) trajectory quite well. Wal-Mart, for example, chose an ecological growth strategy in 2005, introducing fuel cells for electricity at many of its centers, selling goods made from recycled materials, and adopting solar panels at numerous outlets in the United States and abroad— vital to its cosmetic shift from a negative to positive enterprise in local communities.[25] Such faddish change, however, brings forth little progressive substance: the carbon footprint remains just as destructive as "greening" innovations are easily countered by other factors such as increased overall growth and basically eco-*unfriendly* practices in all other phases of business activity around the globe.

Avenues of such reform today are limited by a complex set of structural and ideological conditions. Political institutions, riddled with special interests and labyrinthine processes, suffer impasse at a time when ruling elites have become more insulated, provincial, and detached from the global and long-term consequences of their agendas. An ideological void outside the liberal consensus permits a wide range of escapist and delusional responses, Hedges adding that "The inability of liberals and the power elite to address our reality leaves the disenfranchised open to the manipulation of demagogues."[26] Manipulation is typically the province of the corporate media, which, aside from its nonstop commercial advertising, is inclined to spin tales of foreign threats, socialist takeovers, and nightmarish government deficits to mask the harsh realities of growing unemployment, home foreclosures, eroded social programs, collapsing infrastructure, and environmental decline. Liberals are obligated to the same corporate-military interests as conservatives, each buying into the legitimating myths of a deteriorating social order ruled by ever-defiant elites. History shows that oligarchical power allows no easy departure from its relatively closed (though seemingly open) system of domination. It is true that liberals take environmental problems more seriously than conservatives, though in recent years liberal initiatives have gained little headway, as with the 2010 proposed global-warming legislation that was blocked in a Democratically controlled Senate. More ambitious liberal overtures like those of Gore, Friedman, and Brown are destined to face equally imposing obstacles.

So long as corporations rule, liberal environmentalism cannot escape the paradigm of growth and profits: capitalism will fight any constraints on its freewheeling *modus operandi*. Some costs of reforms can be thrown onto taxpayers and consumers but, since this can be politically difficult, the more feasible alternative is to oppose change. As Immanuel Wallerstein observes, "The political economy of the current situation is that historical capitalism is in fact in crisis precisely because it cannot find reasonable solutions to its current dilemmas, of which the inability to contain ecological destruction is a major one, if not the only one."[27] This impasse raises deeper questions about the exercise of corporate power: beneath Enlightenment pretenses of human progress through ceaseless technological and industrial expansion, humanity faces a world of poverty, waste, violence, and destruction. Elites respond by moving full-speed ahead, doing everything possible to exploit the world's land, water, natural resources, local communities, and labor, transforming everything into useful but disposable commodities. Globalization endows multinational corporations with new scope, dependent on continued exploitation of fossil fuels. Their contempt for planetary life-support systems recognizes few if any limits, the basis of their crusade behind scientific denial and political obfuscation. Moreover, as the global system is relatively planless and Darwinian, devoid of effective regulations, it is less amenable to Keynesian or social democratic interventions of the sort historically adopted in Europe. Where people and nature become simultaneously objectified commodities, as malleable units of production, hopes for sustainable development vanish. A colossal engine of wealth accumulation, world capitalism views the natural environment as a bottomless resource pool to be tamed, exploited, and marketed, with social and ecological values rarely entering the profit-driven calculations of those who manage the system. In Joel Kovel's words: "The very fluidity sought by capital imposes ever greater demands that profits be made right away or sooner. This is the main reason why nothing substantial will be done about global warming under the present regime."[28] Not only is the economic regimen alienated from nature, so too are the people (managers, workers, consumers, media propagandists, etc.) involved at all levels: the restless drive to master nature—a legacy of both the Enlightenment and capitalism—ensures economic collapse as well as ecological disaster. With the ecosystem so thoroughly instrumentalized, the path toward destruction surely follows where, as Theodore Adorno and Max Horkheimer write, the opposites of progress and barbarism, life and death, heaven and hell dialectically but precariously hang together.[29] Despite all outward appearances of rationality, the Enlightenment valorization of technological rationality, industrial growth, and material abundance gives rise to ecological disaster in the midst of "development" and "progress." The predictable result, suggest Adorno and Horkheimer, is "as totalitarian as any system."[30] Yet "totalitarianism" in this case signals more a downward spiral toward breakdown, chaos, and warfare than a system of monolithic or impenetrable rule. Global entropy reflects nothing so much as the increasingly dark side of the Enlightenment project.

The problem with liberal fetishism of "the market"—in reality nothing but a tribute to unprecedented corporate power—is that this dark side of the Enlightenment is ideologically suppressed and hidden at a time when global capitalism exerts greater economic and political leverage than ever. Liberals seem to believe that reforms secured within and against this insidious apparatus of domination can somehow, magically, reverse the global crisis.

Technological Remedies?

If liberals see market solutions as one axis to reversing the ecological crisis, ultimate success requires a widespread shift toward alternative technology. The departure from a carbon-based economy depends on the widespread switch to clean, renewable energy sources (solar, wind, geothermal, plant-based) combined with more efficient use of oil, coal, and natural gas. Conservation technologies are expected to facilitate the greening process. Transportation grids, homes, office buildings, factories, schools, and the military would be restructured to meet renewable energy priorities. Supercomputers and other new-technology options would significantly help transform and regulate the field of energy consumption.

Friedman argues that green power is the wave of the future and that the United States—given its supremacy, however shaky, within the world economy—is perfectly situated to be a driving force behind that "mother of all markets," green technology.[31] In carrying out this epic project, the United States would create a new global model, allowing it to assume economic, technological, and moral leadership.[32] Fusing market ingenuity and technological innovation, a renovated world capitalism could be genuinely transformative to the extent it phases out fossil fuels and replaces them with greener, more efficient, less costly energy sources. In Friedman's view, such greening need not conflict with economic development for the system "can grow without limits."[33] A massive embrace of renewable technology, including electric vehicles, would take the world closer to net-zero emissions, rendering global warming is solvable problem. To be sure, such a "revolution" would necessitate dramatic lifestyle changes—reduced electricity in homes and offices, water conservation, limited use of autos, recycling of consumer goods, and so forth. Friedman also envisions an expanded role for nuclear power and greater reliance on "cleaner" coal. To coordinate this transition he proposes an all-powerful "energy Internet" capable of organizing and regulating large-scale energy systems like utility grids and wind farms.[34] The federal government would create a "nationwide renewable energy mandate" along lines of wartime planning and mobilization.[35]

The liberal stratagems of Gore, Brown, and Friedman all focus on market, technological, and lifestyle changes. Both Friedman and Gore want the United States to reclaim its "moral authority" as international leader by developing green technology within a framework that recalls the Manhattan Project or Marshall Plan which, of course, depended heavily on government initiative, planning, and money. Gore refers to a "Great Transformation," calling forth an epic historical challenge.[36] And like Friedman, Gore believes a qualitative shift toward renewable energy is compatible with continued economic growth, perhaps even with a boost. Gore also views computer technology itself as a dynamic greening agent: just as the Internet holds out great hopes for democratic renewal, it offers a wealth of technical solutions in the fight against global warming and in efforts to design a green economy.[37]

Brown's proposals, informed by many years of environmental research and analysis, follow these contours but places stronger emphasis on lifestyle changes than Friedman or Gore. His attitude toward capitalism appears somewhat more jaundiced, however, probably tied to the conviction that a more radical strategy is needed to meet the ecological challenge, writing, "We know from our analysis of climate change, from the accelerating deterioration of the economy's ecological supports, and from our projections of future resource use that the Western economic model—the fossil-fuel based, automobile-centered, throwaway economy—will not last much longer. We need to build a new economy, one that will be powered by renewable sources of energy...."[38] Yet the "new economy," for Brown, depends more on technological innovation than on corporate restructuring. Brown does propose a stiff carbon tax as one means of coaxing economies toward alternative methods. Like Friedman and Gore, he believes it would take a form of wartime planning to bring the United States (and indeed the world) into the dawning green era. It is generally agreed that resources are more than sufficient: "We have the technologies, economic instruments, and financial resources to do this," notes Brown. Further, "The United States, the wealthiest society that has ever existed, has the resources to lead in this effort."[39] A new, more progressive tax code is needed to render "the market [more] ecologically honest," enabling the United States "to launch a crash program to shift to plug-in and all-electric hybrid cars while simultaneously investing in thousands of wind farms...."[40] For Brown, a Herculean project to reverse global warming is indeed possible (and also profitable) in coming decades—within the confines of liberal reformism.

Such unwavering faith in the blessings of technology might coincide with demographic changes said to be sweeping the United States and other countries. Morley Winograd and Michael Hais, in *The Millenial Makeover*, argue that generational changes in the United States are tied to the spread of Internet-based communications, greater openness to novelty, and belief in technological innovation of the sort required to solve urgent social and ecological problems. This "tsunami of change" rooted in the new generation signifies nothing short of a "radical shift in power" which the authors believe was vividly demonstrated by Obama's 2008 election to the White House.[41] For contemporary youth new ideas appear to flow freely, advanced technology is easily grasped as a mode of communication and empowerment, change is readily accepted, and environmentalism is largely taken for granted. Young people today are more likely to recycle used goods, ride bicycles, drive hybrid cars, live modestly, and eat lower on the food chain that their elders. A "social-networking" population is more open to diverse lifestyles, experimentation, and social novelty, suggesting a potential "tectonic shift" in American politics marked by rebellion against conventional thinking and renewed faith in governmental action.[42] The Obama presidency, energized by the Internet generation and unprecedented online fund-raising, is upheld as the most dramatic expression of this Millennial shift: the White House has pushed for increased high-speed Internet spending, more light-rail systems, and green-technology investment linked to creating jobs, rebuilding the public infrastructure, and giving the United States enhanced leverage in the global economy, Obama noting, "If we want the next technological breakthrough that leads to the next Intel to happen here in the United States, not in China, not in Germany, then we have to invest in American research and technology...."[43]

The modern quest for technological "greening" takes many forms—electric cars, solar panels for electricity, "cleaning" of fossil fuels, wind turbines, smart-grid systems for cities, eco-friendly buildings, and so forth. Indeed, U.S. investment in alternative energy sources has already taken off at the start of the twenty-first century, giving rise to a growth industry for thousands of companies, large and small. New-tech firms like Airborne Wind Turbine, the Cleantech Group, and Greentech Media often work in partnership with universities doing research on green initiatives. Such innovation is surely necessary for any ecological revolution, but whether all those enterprises in tandem can furnish more than 10 percent of total American energy needs is questionable. Whether smaller firms can ultimately resist being swallowed up by such corporate behemoths as ExxonMobil, Bechtel, and PG&E is equally problematic. The increasing popularity of green buildings in the United States and Europe holds great promise, though the issue of *net-energy* utilized (considering fossil fuels used in construction, maintenance, etc.) still must be addressed. The Bullitt Center in Seattle stands as a model of alternative architecture and urban greening—a massive edifice built of toxin-free materials that can generate its own power, uses its own rainwater, and composts its own sewage. At the Center tenants are required to use office equipment that automatically shuts down when idle. Such eco-designed buildings could become the wave of the future, reducing carbon emissions while moving toward net zero-energy consumption. This obviously goes beyond superficial corporate "greenwashing," but whether the vast majority of *existing* structures (in the United States and the world) can be similarly greened is highly uncertain.

Reversing the global crisis will demand a creative mixture of renewable energy sources, as technological breakthroughs allow for elevated problem-solving and social change. While liberals fondly embrace alternative technology, few believe technology alone can solve a deep and multifaceted ecological threat; no such fixes are available. As noted, other change agencies are identified: market restructuring, government planning, tax policies, lifestyle shifts. Yet, insofar as liberals are obsessed with renewable *energy* sources, their "solution" does in fact gravitate toward technological remedies, perhaps even miracles. Technological solutions revolve around a peculiarly instrumental solution to complex, intractable social problems. For decades, in fact, environmental challenges have invited technological responses—renewable energy to fight climate change, a "green revolution" to solve agricultural shortages, water desalination schemes, a shift toward electric or hybrid cars, green buildings, and so forth. It would be foolhardy to discount such innovations, yet since they fail to address broader questions of production and consumption, they inevitably fall short of an ecological politics. A singular technological focus ignores the social and *political* totality, which cannot be reduced to manageable technical (or market) solutions, despite temptations to do so. The Manhattan Project legacy is naturally appealing but the task of building a doomsday weapon, however daunting, should not be confused with the project of transforming the world economy consistent with developmental sustainability. We already know that strictly (or primarily) technological fixes thrown at such time-honored issues as war, poverty, and disease have had little if any impact—a recognition that ought to elicit humility as we face the worst planetary threat yet. This is no purely theoretical critique: the work of Friedman, Gore, and Brown, while informed by a sense of urgency, ultimately avoids or finesses too many dimensions of the global crisis to provide a strategy for ecological transformation. The main problems with liberal-reformism can be summarized as follows:

1. The dominant forms of corporate power—transnational enterprises, industrial giants, banking, agribusiness, media conglomerates—will have to be overturned with leading sectors of the economy socialized, transferred into public hands, and eventually broken down and democratized. (A fundamental dimension of change fully ignored by liberals.)
2. Key sectors of the public infrastructure must be renewed, expanded, and rendered more sustainable, galvanized by a gradual exit from carbon energy sources and a shift toward universal social programs and services. (At best only partially addressed.)
3. A process of de-suburbanization must be set in motion, facilitated by localized delivery of goods and services, giving rise to a milieu in which people can work and be socially engaged closer to home along lines of a new urbanism. There can be no immediate or total "exit" from sprawling and dysfunctional urban and suburban population centers; rather the point is to transform these areas of life to fit more human-scaled, livable, sustainable arrangements. (A necessary transformative process, so far largely ignored.)
4. A gradual but ambitious conversion of resources and public spending from military to social priorities will have to occur, especially in the United States with its gargantuan warfare state. This will free

revenues for technological innovation, infrastructural renewal, and sustainable development, so that reductions in economic growth will be consistent with higher standards for the general population. This is particularly urgent insofar as the U.S. armed forces alone contribute nearly 2 percent of total worldwide carbon emissions. (A perspective that liberals fully ignore.)
5. Agricultural production must undergo a qualitative shift from animal-centered farming and fast foods toward more resource-efficient plant-based farming, maximizing global food resources and health outcomes while reducing toxic wastes, runoffs, and other pollution including carbon emissions. (Conventionally ignored, though at least peripherally addressed in Brown's Plan B.)
6. Efforts must be undertaken to limit population growth, which if left unchecked will soon overwhelm the earth's carrying capacity. Policies geared to education and taxation can be adopted on a country-by-country basis, depending on a variety of demographic factors at work. (An issue typically dismissed or sidestepped, although Brown does place it at the heart of his sustainability program.)
7. Government policies must be directed toward modern, efficient, sustainable public transportation in urban areas—and between urban areas—with the goal of diminishing reliance on personal cars marked by a radical downsizing of the auto culture. (Among liberals this is at best partially addressed.)
8. A corollary to the first point, corporate and "private" monies, including lobbies, must be decoupled from the political process, allowing elections and legislative activity to proceed more democratically in the absence of big-business agendas that, as we have seen, run directly counter to ecological imperatives. (A crucial imperative that is completely ignored.)

The overarching problem here is that liberals, of whatever stripe, remain thoroughly attached to the corporate-state as they propose to reform and improve it through wider use of green technologies, new market incentives, and sustainable lifestyles—supported, to be sure, by an active public sector. Unfortunately, the main pillars of the corporate-state would remain fully intact: huge business empires, the warfare state, Wall Street, agribusiness, a political order riddled with predatory interests—fueled by a perpetual growth machine. New technologies would permit a continuous "greening" of the state-capitalist order, though without an ecological model rooted in socialized production and consumption, drastic reduction of fossil fuels, and more democratic, egalitarian forms of social organization. Wedded to the prevailing arrangements, liberals cannot take the system beyond its unsustainable logic. No liberal program can pursue reduced growth as part of an ecological model, Richard Heinberg noting that "The energy conundrum is thus intimately tied to the fact that we anticipate perpetual growth within a finite system."[44] For the United States, this means 5 percent of the world's population will continue to utilize (or at least *strive* to utilize) 25 percent of available global resources, based on endless replication of dysfunctional consumption habits linked to cars, suburbia, shopping malls, oversized homes, and fast foods not to mention Pentagon allocations.

Still, faith in liberal "alternatives" of the sort presented by Gore, Friedman, and Brown dies hard, in part owing to the apparently limited demands of such reformism on both the ruling interests and ordinary people. Their technological overtures, for example, promise sustainable energy well into the future, encouraging Americans to indulge their addiction to cheap oil and everything it provides. In a context of impending peak oil and worsening effects of climate change, however, such promise turns into a mirage. Thus Michael Ruppert notes, "The truth is there is no alternative energy, or combination of alternative energies, that will permit current consumption and lifestyles to continue—let alone provide for the compound growth we are wedded to in the current economic paradigm."[45] If most Americans remain in a general state of denial, more enlightened actors gravitate toward quick, easy, and relatively cheap "solutions," understandably seductive given the long U.S. history of inventions with far-reaching social consequences: autos, film, radio, TV, computers, airplanes, antibiotics, and nuclear power among many. Popular faith in salvation through "green energy" thus has a compelling logic, even as ruling elites strive (against increasing odds) to manage economic affairs just as they have for the past century or more. The quest to replace cheap oil takes many forms—solar and wind power, geothermal energy, hydrogen, tar sands, plant sources, nuclear energy, "clean" coal. Yet, as Rupert and others have persuasively shown, when the net-energy calculation is fully taken into account, all such forms combined cannot provide a workable alternative to cheap and readily accessible petroleum under existing patterns of production and consumption.[46]

A shift toward clean technologies is of course unavoidable and indeed already an established fact, but its long-term potential is far more limited than optimistic scenarios presume. Solar and wind sources hold great promise, but they are expensive, restricted in usage, and typically so energy dependent that net benefits are considerably less than usually projected. The notion of a "hydrogen economy," faddishly embraced by some environmentalists and politicians, is a pipedream, requiring even more energy than it can produce. Biomass schemes like ethanol are not only extremely polluting but consume enormous oil supplies including petroleum for fertilizers, pesticides, and delivery. Kunstler notes that such alternatives "are predicated entirely on the assumption of an underlying fossil-fuel platform, especially in terms of agricultural waste products such as cornstocks grown under an industrial agriculture regime using massive petroleum and natural gas 'inputs' for artificial manufactured fertilizers, harvesting, and transport." In reality "the amount of petroleum and natural gas needed to produce the corn to make the ethanol would more than cancel any benefit from using a supposed non-fossil fuel."[47] Synthetic oil likewise has limited promise, whatever its source. For example, tar sand deposits in Canada are massive, but Heinberg points out that "the rate of extraction of that resource will always be limited by the fact that the producing process is—and will inevitably continue to be—expensive from both a financial and energy standpoint, as well as being environmentally disastrous."[48] Alberta's output in 2010 was about one million barrels daily—a tiny percentage of U.S. total consumption—but the mining projects require up to 700 cubic feet of natural gas to produce a single barrel of synthetic crude, while destroying many square miles of natural habitat. Similar problems accompany coal-derived synthetic oil, which too is expensive, restricted in utility, and ecologically ruinous, Kunstler noting "The only plausible application of coal-derived liquid synfuels will be in the military, and even that is debatable."[49]

Anticipating the onset of peak oil, many look to coal as a "clean" alternative—a scheme pushed strongly by Obama. In 2010 coal provided nearly 30 percent of world energy needs and generated 40 percent of its electricity, though with horrendous environmental impact. In the United States dozens of coal-fired plants now operate, with dozens more under construction. The problem is that even where coal is processed to reduce greenhouse emissions, coal-mining enterprises generate equivalent amounts of toxic waste and land despoiliation. And most coal-generated energy today comes from what is best described as dirty plants. As coal becomes more difficult to extract, moreover, it ends up more expensive and land destroying even while the industry promotes it as a post-carbon energy source. Former TVA head Dan Becker contends, based on lengthy field experience, that "There is no such thing as clean coal and there never will be. It's an oxymoron."⁵⁰ Rupert is even more scathing, insisting that "Clean coal is one of the biggest lies in human history."⁵¹

Could nuclear power then be the technology needed to fill the energy void and save the world from the twin threats of peak oil and global warming? In 2011 the United States derived 20 percent of its electricity from 104 nuclear power reactors, although—because of expense and politics—no new plants were entered into commercial operation after 1996. Some countries (including France and Japan) gain considerably more of their electricity from nuclear energy. Greater reliance on nuclear power could possibly help the United States (and the world) transition to a post-carbon economy, especially if uranium deposits hold out for several more decades. Energy from nuclear fission easily dwarfs that from other sources, but massive obstacles are well known and inescapable, more visibly so in the wake of the Japanese nuclear disaster following the March 2011 earthquake and tsunami: accident scenarios, disposal of spent fuel, radiation emissions, and of course nearly prohibitive cost. In the United States, as elsewhere, nuclear plants are heavily subsidized by taxpayers. A deeper conundrum is raised by Helen Caldicott, who suggests that in the long run the nuclear option will yield little if any energy, writing, "Over several decades, as the concentration of available uranium ore declines, more fossil fuels will be required to extract that ore from less-concentrated ore veins. Within 10 to 20 years, nuclear reactors will produce no net energy because of the massive amounts of fossil fuel that will be necessary to mine and to enrich the remaining poor grades of uranium."⁵² (Caldicott more recently argued that the global nuclear industry would probably end up moribund, in part owing to the 2011 Japan catastrophe.) Further, as Kunstler points out, relatively few human activities will be supported by nuclear power even at its most robust—and those pointedly do not include transportation. It will compensate little for precipitous losses incurred by fossil-fuel depletion.⁵³

At a time when Republicans still worship fossil fuels, looking for quick solutions in stepped-up offshore drilling, oil shale, natural gas, and "clean" coal. Democrats push alternative energy sources as an even quicker fix—especially in states like California, where solar power is being heralded as the great remedy for global warming. By 2011, riding an ambitious $3.3 billion initiate, California embarked on a solar campaign hoping to install solar panels on at least three million homes by 2016, an eminently achievable goal with prices for photovoltaic cells dropping. The state planned to establish an energy system that would provide 16 percent of all electricity by 2030, which on the surface looks to be a great leap forward. Appearance, however, are greatly deceiving: solar gains will do virtually nothing to reduce California dependence on fossil fuels since overall energy demands will continue to skyrocket with the steady growth of population growth, meat-centered agribusiness, cars, and other vehicles—little of which will be satisfied by a turn toward renewable sources. In fact projections indicate that, by 2035, solar energy will supply only 1 percent of total California demands—hardly the realm of quick technological remedies.[54]

Frustrations in the wake of extremely disappointing fates for electric cars offer yet another illustration of misplaced technological optimism. A Think City auto plant near Elkhard, Indiana was expected to produce some 60,000 electric vehicles annually by 2013, but as of early 2012 the factory was essentially idle, with just two workers hired instead of the anticipated 400. After receiving millions of dollars in government aid and tax breaks, both the auto maker and battery suppliers were force to file for bankruptcy; public demand for relatively expensive, weakly performing electric cars was virtually nonexistent. Vice President Joe Biden, referring to the dream of a gasoline-free future, had framed Think City as a vital project in the remaking of America after President Obama visited the site on three occasions—but the plant was never able to hire more than 25 employees.

American liberalism has long been associated with an ethos of technological optimism, part of the oil age with its dramatic advances in machinery, electronics, transportation, agriculture, and of course the military. Technology has been regarded as something of a miraculous force in the service of economic growth and political power, one reason government and business leaders approach questions of oil depletion and global warming as merely another technological challenge. Evidence already suggests, however, that even when strictly applied to matters of sustainable energy, purely technological answers do not exist—more emphatically so when it comes to the global crisis. The seductive nature of fixes sidesteps necessary, but always messy and difficult, requirements of an ecological politics, just as it fails to address hegemonic corporate and banking interests that tenaciously work to defeat alternatives to the corporate-growth economy. It also refuses to acknowledge, in Kunstler's words, that "virtually all the economic relationships among persons, nations, institutions, and things that we have taken for granted as permanent will be radically changed during the Long Emergency," as life becomes "intensely and increasingly local."[55]

LIBERALISM EXHAUSTED

Pursuit of liberal reforms, even where motivated by the best intentions, cannot transcend restraints of corporate-state power that, for the United States, are further enforced by the warfare system and Empire. Aspirations for change might be keenly embraced, but an efficacious *strategy* pointing a way out of the global crisis is lacking. Liberal environmentalists like Gore, Friedman, and Brown wax eloquently about the need for a "green revolution" through "market restructuring," clean technology, and altered lifestyles, but this "revolution" is bound to existing political and economic arrangements, guaranteeing impasse. These advocates call forth a "radical shift" but inevitably settle for a moderate reformism that retreats in the face of a corporate stranglehold over American public life. Brown proclaims an opposition to business-as-usual spurred by an emergency national mobilization, but never advances a coherent strategy for its realization. At the start of the twenty-first century, as the ecological crisis deepens and corporate power expands, the liberal-reform model encounters heightened obstacles. An ecological politics will surely have to advance reforms, big and small, but political-strategic quagmire occurs when the liberal path is framed as the *only* route to oppositional politics.

Liberalism, unfortunately, cedes both institutional and ideological terrain to the ruling interests as its *modus operandi* depends on normal politics that leaves those interests fully intact. Its ideology has long been wedded to the vagaries of corporate power and, for the most part, its institutional politics has followed suit. European-style social democracy constitutes the political ceiling of liberalism, historically adopting strong Keynesian policies to socialize the economy and restrain corporate power—a *social* Keynesianism in contrast to the American military variant. Even as social democracy enlarges the scope of progressive regulations, laws, policies, and treaties, however, it is scarcely better positioned to face the strategic imperatives of reversing the ecological threat than mainstream American liberalism. The more conservative response—looking for a return to mythical laissez-faire capitalism—is no solution, as its support of unfettered corporate power portends environmental as well as economic catastrophe. That leaves three generic strategic modalities: some form of Leninst vanguardism (Jacobin revolution from above), a spontaneist process of anarchist insurgency, or a broadly radical politics along lines of the Green synthesis of popular movements and electoral/legislative politics anticipating full-scale transformation of the economy, government, and culture. These strategies of change share in common one broad goal: overturning of the corporate-state power structure. Twentieth-century experience demonstrates that Leninism, where successful, leads eventually to bureaucratic centralism while anarchism, for its part, has always lacked political-strategic articulation—that is, a capacity to conquer and manage state power. Neither of these approaches is compatible with a politics of *democratic* transformation. The Green synthesis—in its deepest anti-systemic expression—is the most fruitful strategy for reversing the crisis, consistent with ecological politics and sustainable development on a global scale.

The Green model departs fundamentally from liberal environmentalism as it embraces (1) socialization of production and banking, laying the groundwork for a whittling away of corporate power; (2) limits to growth rooted in a steady state economy oriented to post-carbon, sustainable priorities; (3) a move toward demilitarization, with defense functions shifted exclusively toward domestic boundaries and resources directed away from the warfare state, into social investment; (4) de-suburbanization connected to planning for a localized "new urbanism" and human-scaled living arrangements; (5) a shift from auto-centered to public transportation systems, within cities and between cities; (6) large-scale funding of alternative technologies; (7) agricultural renewal involving change from agribusiness to local forms with commensurate de-emphasis on animal farming and fast foods; (8) vastly increased public resources devoted to collective modes of consumption such as health care, transportation, housing, education, and child care.[56] Liberal environmentalism has advanced just one of these objectives with any consistency—alternative technologies—the limited adoption of which turns out to be compatible with (indeed profitable for) corporate interests. A radical green model based on the above program, however, is not likely to make much headway within the corporate state or Washington Beltway. Its ecological model is shaped by a post-liberal ideology and given strategic definition by anti-system forms of organization, leadership, and methods along with immersion in social movements—approximating the radical side of European Greens at their peak in the early 1980s.

Liberal politics in the United States has achieved its fullest expression mainly within the Democratic Party, which—like the Republicans—relies heavily on corporate funding for its electoral and legislative work. Despite contrasts in rhetoric and style—combined with certain social priorities—both parties generally support the elite consensus on issues related to corporate power, Wall Street, the military, and foreign policy. (Here the Tea Party is not as far removed from mainstream thinking as is commonly believed.) Given the decisive role of lobbies, think tanks, foundations, and the media in American political life, the extent to which they favor Republicans gives that party a clear advantage in steering political discourse around crucial policies: military spending, the war on terrorism, taxes, deficits, the environment. Coincidentally, the space for dissent is narrowed as change-oriented movements encounter new roadblocks; as corporate power expands, centers of independent activity and citizen empowerment tend to shrink. New Deal politics, moreover, with its broad acceptance of ambitious public agendas and governmental activism, has been under sustained (often effective) assault since the Reagan era, extending to the Tea Party insurgency. Public institutions have become less vibrant, less solvent, less efficient, and ultimately less accountable. A political system and mass media that celebrate democracy and freedom remain mired in largely irrelevant, contrived debates, patriotic sloganeering, and ideological platitudes—those platitudes embedded in a long history of American exceptionalism in which anti-system views are regarded as "un-American" or just plain crazy. At present even populist discourses seem to favor rightwing narratives that are mostly consistent with elite priorities. Hedges writes that the "The inability of liberals and the power elite to address our reality leaves the disenfranchised open to the manipulation of demagogues."[57] Sadly, American politics today is so greatly distanced from the threat posed by the global ecological crisis that it cannot serve as an arena of meaningful reform—much less radical transformation—which is the unfortunate fate of liberal "solutions" to the ecological crisis.

Chapter 5

Struggle For an Ecological Politics

Several decades of environmentalism in the United States and other advanced capitalist societies have left the world considerably short of an ecological radicalism needed to confront the global crisis, despite important reforms achieved here and there. Liberal agendas have failed on two fronts, offering neither an alternative mode of economic development nor a workable political strategy, while corporate expansion reshapes the global economy, politics, finance, culture, social life, and the natural habitat. Reforms have done little to ameliorate a system that develops in harsh conflict with nature, labor, consumers, and local communities. As ecological disaster lurks, the ruling interests do everything possible to loosen controls over their power, resist change, and close off democratic inputs as they bring the world heightened geopolitical conflict, environmental ruin, militarism, and war. That such a powerful and well-defended system could be transformed from within its own fortress, either domestically or globally, seems increasingly far-fetched. A "greening" of world capitalism is something of a liberal fantasy: environmental sustainability is logically incompatible with a regimen of endless accumulation, growth, and exploitation. The Washington Beltway, like other bastions of power, has been unfriendly to even modest reforms as shown by the dismal fate of climate change legislation in 2010. Liberal environmentalism, itself beholden to corporate largesse, has capitulated to energy and utility industries, fearful of directly taking on the big polluters. Johann Hari comments, "They [environmentalists] take money, and in turn they offer praise, even when the money comes from companies causing environmental devastation."[1] Members of Congress are indebted to contributions from Big Oil, Big Coal, and Big Utilities, meaning serious climate change reforms face enormous obstacles in that arena. Hari adds, "After decades of slowly creeping corporate corruption, some of the biggest environmental groups have remade themselves in the image of their corporate backers."[2]

APOCALYPTIC SCENARIOS

Recognizing the intractable problems of liberalism, some writers, commentators, and activists (though few politicians) embrace more radical solutions to derail catastrophe. In this view, barbarism is the fate of humanity unless an ecological politics can shape the global agenda, as reforms alone cannot suffice to interrupt the downward slide. Such post-liberals include Richard Heinberg (*Power Down* and other books), Bill McKibben (*Eaarth,* his most recent), James Howard Kunstler (*The Long Emergency*), and Michael Ruppert (*Confronting Collapse*), their work helping to spread whatever limited ecological awareness presently exists in the industrialized societies.

Heinberg sets forth perhaps the most ardently anti-system outlook among these critics, focusing on the combined threats of climate change, peak oil, resource depletion, population pressures, and economic crisis. Heinberg devotes special attention to the problem of oil depletion, which he expects to "peak" by 2016 if not sooner. The world will soon exit the era of cheap, abundant fossil fuels, he argues, with global demand beginning to outstrip supply, calling into question present levels of manufacturing and transportation as well as food production. For the United States, the threat is exacerbated by a rapid decline of domestic oil reserves, which could vanish before 2030. Dismissing nuclear power and "clean" coal as viable energy alternatives, Heinberg is pessimistic about technological solutions that he believes cannot fully substitute for fossil fuels in transportation, food production, and general electricity needs. Historically an engine of the industrial revolution, hydrocarbons have morphed into a great curse, which for Heinberg demands a thorough rethinking of established modes of production and consumption. Departing from the liberal approach, Heinberg insists that corporate and military power must be targeted since Americans remain trapped in a perpetual growth machine devouring one-quarter of the earth's resources. Unlimited consumption is premised on infinite natural resources that fuel a culture of cars, malls, fast food, and suburbia. Thus, "It is our reluctance as a species to undertake demand-side solutions to the ecological dilemma—not merely our inability to find a suitable substitute for oil—that is leading us toward collapse."³ Collapse is the unavoidable scenario ahead if radical action is not undertaken, and soon. Heinberg foresees a Hobbesian world of chaos, violence, and warfare as both nations and corporations intensify their competition over shrinking resources across the planet. For the United States, disruptions will hit every realm of human activity, eventually giving rise to any number of morbid ideological responses. Collective denial is bound to accompany social breakdown and political despair, hastened when the search for quick, pain-free solutions proves illusory and elite attempts to maintain their power, wealth, and institutional leverage intensify. Meanwhile, alienated mass publics can be expected to feel increasingly detached and disempowered.

Other apocalyptic thinkers generally share Heinberg's dystopic view of the future: McKibben, Kunstler, and Rupert articulate, to varying degrees, Heinberg's anticipation of a global collapse marked by deepening chaos and violence. They agree that the illusory search for market-based and technological fixes, while perhaps comforting in the short term, cannot sustain existing patterns of economic growth, militarism, agricultural waste, and suburban lifestyles all dependent on the oil economy. None has faith in a "green revolution" that would further rationalize corporate power while extending old habits and addictions made possible by cheap, abundant fossil fuels. They share a jaundiced view of American politics as immobilized by the power of vested interests, a labyrinthine state system, and permanent war economy. Through all this, meanwhile, Americans remain confined to what Kunstler calls a "consensus trance," vulnerable to myriad forms of collective escape, ideological delusions, and political fantasies that insulate them from the real global challenge.⁴

The spread of a virulent pro-corporate outlook in the United States further harms prospects for environmental change that depends on public goods over private interests, sustainable development over endless growth, waste, and destruction. To question the ethos of corporate profit making is widely seen as un-American, irresponsible, subversive, while to criticize economic "growth" is to open the door to developmental alternatives the public cannot even *imagine*.⁵ Where problems are in fact noted, answers typically involve renewed emphasis on growth, profits, and deregulation that the public is told will solve crucial problems, a reassuring elixir as people are happy to believe in trouble-free solutions. Few Americans, after all, are ready to give up lifestyles embedded in the car culture, suburbia, private consumption, meat-based diets, and global military power. Such views are reinforced by a long history of American exceptionalism that confers

a special national mission and identity. The world has always been rich in natural resources for the United States to exploit, backed by the military force to do so, with few restraints. The threats of peak oil, depleting resources, and climate change are taken as just more unwelcome bad news, likely overblown and in any case just temporary nuisances. The tantalizing question is how long ideological escapism of this sort can persist as the global crisis escalates. Kunstler argues, "Behind peak [oil], things unravel and the center does not hold. Beyond peak, all bets are off about civilization's future."[6] As matters worsen, Kunstler expects a "bunker mentality" to take hold among Americans, indulging in the "mass delusion" that life can go on merrily as before. Convinced that the system of doing business is natural, fixed, and unchangeable (perhaps even God ordained) people readily cling to outmoded myths to the bitter end. For Heinberg, McKibben, Kunstler, and Rupert the crisis possesses its own irresistible momentum, with no comfortable exit. Heinberg sees a downward spiral in which people are increasingly demoralized, bereft of collective self-activity. McKibben, for his part, anticipates continuing global cataclysm ahead. Recognizing limits to conventional reformism and the need to jettison the corporate-growth model, he looks to more radical though still politically fragmented solutions: localized agriculture, green technology, lifestyle shifts, human-scale communities, more restrained consumption. Typical of the apocalyptic sensibility, McKibben never furnishes a political strategy for translating specific ideas into broader, more integrated structural changes.[7]

Kunstler's pessimism is more extreme yet, heightened by a cynical view of American public opinion that he sees mired in collective denial and ideological escapism. Like Heinberg, he rejects strictly market and/or technological solutions as ultimately self-deceptive. He anticipates a chaotic breakdown accompanied by a mad scramble, individually, societally, and globally, for declining natural resources like food, water, land, and oil. The fate of industrial society is entropy followed by more entropy—a descent into anarchy and violence most likely to be met with political *immobilisme*. Kunstler, understandably, places less faith than the liberals in lifestyle changes along lines of consumption downsizing, which can be little more than cosmetic initiatives, writing, "The Long Emergency will cause unprecedented social and economic dislocation, and the outcome may be a world we would barely recognize."[8] Kunstler's dystopic outlook dashes hopes for political action to counter imminent collapse. Ruppert's narrative is hardly more reassuring, sharing with Kunstler an understandably dark view of American political life. The global crisis is so far beyond the reach of conventional politics that familiar approaches, including green technology, a shift toward localism, market restructuring, and personal transformations, are bound to fall dramatically short of either "fixing" the system or creating an alternative under severe time pressures. While Rupert expects mounting upheavals in the midst of social and ecological breakdown, he envisions no revolutionary in a context where people are inclined to defend the old ideologies and lifestyles. Like McKibben, Heinberg, Kunstler, and Rupert never so much as suggest the outlines of an alternative political strategy.

IN PURSUIT OF RADICAL ECOLOGY

The struggle for an ecological politics reflects profound disillusionment with the capacity of liberal reformism to furnish alternatives to the present impasse. It recognizes the imperative of overturning a corporate-state power structure that operates as an implacable enemy of nature. Contemporary ecological radicalism is mostly the outgrowth of 1960s new-left movements and counterculture, inspired by the work of Rachel Carson, Murray Bookchin, and Barry Commoner, which galvanized an environmental consciousness leading to the first Earth Day in 1970, rapid spread of academic and media interest, and proliferation of groups that have often survived across the decades. This new ecological sensibility began to raise questions about the destructive side of industrialism, the dangers of a toxic world, and the need to restore balance between society and nature, humans and their environs. Its insurgent, at times utopian, outlook followed a trajectory largely independent of Marxism, dedicated to the ideals of local community, ecological renewal, mutual aid, limits to growth, and generalized opposition to hierarchy and concentrated power. It called forth a rich theoretical legacy grounded in nineteenth century anarchism and utopian socialism, the seminal communitarian ideals of Jean-Jacques Rousseau and Peter Kropotkin, and the work of later critics of industrial modernity like Paul Goodman, Lewis Mumford, and the Frankfurt School (notably Herbert Marcuse). In its most articulate expression, ecological radicalism carries forward an uncompromisingly egalitarian, antiauthoritarian outlook directed against the entire matrix of domination.

Bookchin's theory of social ecology, shaped by classical anarchism and developed through his prolific writings going back to the early 1960s, represents probably the most sophisticated variant of ecological radicalism today. His work is defined by a "dialectical naturalism" in which efforts to transform history and nature, society and environment, are viewed as unfolding simultaneously, eventually giving rise to an organic community—a process driven by local struggles against multiple forms of domination: class, bureaucratic, racial, gender, cultural, ecological. For Bookchin, whose work embellished motifs shared by the new left, a rational ecological order meant full realization of "free nature" through different modes of human self-activity, dependent on renewal of natural relations and spread of new (libertarian) human values. Thus, "Such a change would mean a far-reaching transformation of our prevailing mentality of domination into one of complementarity, in which we would see our role in the natural world as creative, supportive, and deeply appreciative of the needs of nonhuman life."[9] Human and natural worlds would be organically reunited after centuries of harsh opposition and conflict. Bookchin inherited the political radicalism of Marx in his embrace of dialectics and popular struggles to overthrow capitalism, but his anarchist sensibilities take him well beyond Marxism in two important ways: a view toward overturning *all* modes of domination and the struggle for ecological revitalization that is at best only implicit in the Marxist tradition.

Bookchin was a fierce opponent of liberal reformism, believing it would only feed legitimating practices and myths of the system. For him, this was one of the sad legacies of modern environmentalism, for "to 'play by the rules' of the political game means the natural world, including oppressed people within it, always loses something in the end. As long as liberal environmentalism is organized around the social status quo, institutional normalcy, property rights and elite power always prevail over powerlessness."[10] Liberals, in Bookchin's view, suffer from a refusal to see that a capitalism based overwhelmingly on growth and profit making is on a course to destroy the natural world. Writing (in 1983) well before elevated scientific awareness of global warming but still anticipating catastrophe ahead, Bookchin could observe:, "We may well be approaching a crucial juncture in our development that confronts us with a historic choice: whether we will follow an alternative path that yields a humane, rational, and ecological way of life, of a path which will yield the degradation of our species, if not its outright extinction."[11] He argued that "industrially and technologically... we have placed ecological burdens upon our planet that have no precedent in human history."[12] With each increment of destruction and loss, humanity loses a portion of its own essence along with its regard for life (and nature) as such. The problem is not only institutional but extends to the deepest recesses of social life and human consciousness. Thus, "The internalization of hierarchy and domination forms the greatest wound in human development and the most deadly engine for steering us toward human immolation."[13]

Bookchin's view of the ecological crisis—and his implicit approach to its reversal—was informed by an understanding of social and psychological tendencies at the core of the human predicament, writing, "In our discussions of modern ecological and social crises, we tend to ignore a more underlying mentality of domination that humans have used for centuries to justify the domination of each other and, by extension, of nature." Sadly, humans seem dedicated to an "image of a demonic and hostile nature [that] goes back to the Greek world and even earlier."[14] This conception of human impulses goes to the very heart of social ecology, which exposes the "all-encompassing image of an intractable nature that must be tamed by a rational humanity [which] has given us a domineering form of reason, science, and technology—a fragmentation of humanity into hierarchies, classes, state institutions, gender, and ethnic divisions. It has fostered nationalistic hatreds, imperialistic adventures, and a global philosophy of rule that identifies order with dominance and submission."[15] For Bookchin, the "ecological principle" attains "self-actualization" within a society based on community, roundedness of personality, diversity of stimuli and activities, and increasing wealth of experience, a variety of tasks.[16] Following Rousseau and Kropotkin, it is the contemporary task of social ecology to "rupture the association of order with hierarchy."[17]

Bookchin looked to a strategy of "libertarian municipalism" as a means of exiting from the crisis, as a way of subverting the forms of domination, in which the very idea of citizenship, of politics as a domain of community and participatory democracy, has been lost. In the United States "power is thoroughly bureaucratized, centralized, and concentrated into fewer and fewer hands. The power that should be claimed by the people is pre-empted by the state and by semi-monopolistic economic entities. Democracy, far from acquiring a participatory character, becomes purely formal in character... [at odds with] a desire to regain citizenship, to end the degradation of politics into statecraft: the need to revive public life."[18] For Bookchin the solution to the erosion of citizen-based democracy—indeed the ecological crisis—lies in a rebirth of municipalism that has origins in the French and American revolutions and the 1871 Paris Commune. The reclaiming of local community means a turn toward workers' councils and neighborhood assemblies along with confederal relationships between municipalities, in other words: "The need to rescale communities to fit the natural carrying capacity of the regions in which they are located and to create a new balance between town and country—all traditional demands of the great utopian and anarchist thinkers of the last century—have become ecological imperatives today."[19]

Bookchin's theory advances the most radical philosophy and politics yet for confronting the modern crisis. His influence on the new left, counterculture, and new social movements was profound, helping fuel the growth, however stunted, of a "left Green" tendency within environmental politics during the 1980s. It links ecology with other popular struggles as part of a creative, emancipatory vision, upholding faith in the human capacity to reappropriate "first nature" and forge an elevated "second nature" based on reason, democratic planning, and deep citizenship in the service of an ecological order.[20] Yet Bookchin's radicalism—compelling in its prescription for revolutionary change—remains frustratingly vague, even remote on the question of political strategy. One problem lies in Bookchin's easy dismissal of social hierarchy and domination: the question is not whether humans will continue to "dominate" nature—the capacity and even the necessity to do so is undeniable—but precisely what form their intervention will or should take. The corporate-state system is one thing, driven as it is by conquest and exploitation transforming natural habitats into lifeless entities, into reservoirs of wealth and power. But could any social order, even the most ecologically sustainable and democratic, avoid some kind of instrumental, controlling relationship toward nature? Surely the only alternative would be total depopulation of the planet, so that no water, foodstuffs, metals, woods, and paper, for example, could be extracted. The issue here turns on the particular *mode* of human domination, including whether the chosen developmental model is sustainable, roughly consistent with the Earth's carrying capacity and biospheric potential. Bookchin's rejection of all hierarchical forms within human society—including the realm of political governance—suffers from a similar utopianism.

The lack of strategic concreteness in social ecology means that the transformational process remains diffuse, poorly articulated. Bookchin's libertarian municipalism, connecting hundreds of thousands of local communities within a series of confederal arrangements, is laudable enough, but there is the matter of how such a vast, dispersed, and fragmented assemblage of popular formations, even assuming high levels of efficacy, could be coordinated and translated into political reality. Given the pervasive influence of reigning ideological discourses in the industrialized societies, this strategic task would seem to be nearly insurmountable. How could such extremely dispersed social forces hope to overturn the deeply entrenched and unified system of corporate, governmental, and military power that rules the United States. The historical record in hardly encouraging: insurrections and movements aligned with the traditions of anarchism, syndicalism, and council communism have been either crushed by superior power or condemned to political futility. Moreover, even should the project of libertarian municipalism gain headway, the process would inevitably be protracted and uneven—hardly sufficient for a planet veering toward ecological catastrophe. Thus, while social ecology has contributed greatly to an unfolding ecological awareness in the United States and elsewhere, its distinctly *political* value remains marginal.

Standing at odds with social ecology, while simultaneously reacting against both liberalism and Marxism, deep ecology—its influence on green currents powerfully felt since the 1980s—looks to systemic change across the entire scope of human-nature relations, an ecocentric "break with modernity" and industrialism tied to a radical identification with nature. DE shares with social ecology a rejection of all forms of domination but, seeing the urgency of the ecological crisis, looks to natural relations as the privileged site of change. It dismisses liberal environmentalism and the notion of incremental reforms (or "green capitalism") in favor of a more expansive "paradigm change" in consciousness, lifestyles, and values definitive of an evolving human-scale community. DE rejects the Enlightenment legacy *tout court*, urging drastic limits to economic growth, "bioregional" living arrangements, population reduction, self-sustaining agriculture, and unyielding reverence for natural habitats. More fundamentalist DE currents call for a return to preindustrial society or "wild nature" consistent with ecological principles of equality, democracy, peace, spiritualism, and environmental renewal. A founder of DE, George Sessions argues that human self-activity is attainable only through organic unity with the surrounding ecosystem.[21] Some adherents insist that all human interventions in nature are intrinsically destructive and should be minimized where not totally avoided. The modern crisis, according to this extreme view, is surmountable only at a point where the human footprint essentially vanishes—an outlook bringing charges from the left of misanthropic or even fascistic politics. Most within the DE orbit, however, retreat from such dogmatic ecocentrism.

DE theory stresses undying moral obligation to nature and living systems within it, a presumption of biospheric equality conflicting sharply with the requisites of dynamic industrial society: departing from social ecology, DE argues for full-scale transformation of social life and natural relations consistent with the abolition of speciesism, or anthropocentrism. This is no contrived "second nature" but rather progressive adaptation to "first nature," looking to transcend the age-old dualism of society versus nature, humans versus other species, a view more compatible with an animal rights outlook.[22] Its moral stance extends to all facets of the natural world, beyond individual sentient beings to include natural habitats as part of an interconnected ecological system. For DE, radical change demands a qualitative shift in the economy, social structure, political governance, lifestyles, and human consciousness, bringing local communities closer to the surrounding environs.

As with social ecology, the blanket DE rejection of human domination over nature sidesteps a crucial problem—that is, by mistakenly posing the question of domination *tout court* instead of more concretely identifying the character of that domination. In practice, however, DE theorists often lean toward a rather malleable attitude concerning human-nature relations, Arne Naess writing, "My intuition is that the right to live is one and the same for all individuals, whatever the species, but the vital interests of our nearest [i.e., humans] nevertheless have priority."[23] He goes on to defend the use of animals as "resources" for human appropriation,[24] while scattered references throughout DE literature defend the use of animals as food and other commodities. Naess argues that humans should be allowed to intervene in nature "to satisfy vital needs"—seeming more like a sympathizer of corporate power in a fast-food culture than defender of biocentric equality.[25] Lacking any broader social theory, DE leaves moral and political space for humans to continue their destructive modes of production and consumption. Such ecological radicalism would, in fact, never be so "deep" as to interfere with human activities that contribute toward "satisfying vital needs." Conceivably "wild nature" (itself a problematic concept) would remain untrammeled, but in many locales the environment would be vulnerable to merciless abuse on the part of enterprising human actors.

An even greater difficulty with DE is that its promised *exit* from modernity—including the notion of bioregionalism—turns out to be rather abstract and unreachable, a utopian fantasy. Modernity has so thoroughly transformed every realm of the existing world, and has become embedded in social institutions and practices for so many generations, that human attempts to "escape" its complex totality would lead to immediate chaos and breakdown. The abolition of human footprints on the natural world, which no DE theorist has ever concretized beyond individual decisions to relocate closer to the wilderness, ends up as another romantic myth. Biocentric equality, itself a fanciful *human* construct, is far enough removed from any conceivable goal that concrete forms of political action are rendered moot. For ecological thinking DE lacks both specificity and modality of change: "natural" entities from ecosystems to animals, insects, and even trees appear to enjoy the same putative moral standing, however nebulous and subject to qualifications. The extension of moral status across the larger natural landscape seems laudatory enough, but, as Tim Luke observes, such sacralization of nature fails to rise above a vague sense of "moral regeneration" detached from political meaning and strategy.[26] Despite its penetrating, ostensibly radical discourses, therefore, deep ecology offers little guide to tangible ecological change much less a political way out of the crisis.

Whether the growth mania, logic of commodification, and manic exploitation of nature emanates specifically from capitalism or more generally from industrialism, DE offers no "exit" from the crisis: it is left to individuals and small groups to simply depart from modernity, leaving the power structure effectively intact. It offers no incisive critique of the main centers of power, hardly surprising given how DE offers a view of "nature" abstracted from history, society, and politics. As laudably noble defenders of wilderness, therefore, DE theorists and partisans contribute little to organizing for change beyond straightforward lifestyle decisions. Its break with Enlightenment rationality, along with its departure from social ecology and Marxism, unfortunately carries no relevance to actual political outcomes.

From Marx to Ecosocialism

No discussion of the global crisis would be complete without engaging Marxist theory, going back to the seminal work of Marx and Engels and to later twentieth-century Marxist thinkers. The socialist discourses of Marx and others would enter into the formulation of ecological radicalism and leftist tendencies within contemporary Green parties. Some ecological theorists—Chris Williams and John Bellamy Foster among others—rely heavily on Marxism in their analysis of ecological crisis as one manifestation of world capitalism. Williams argues that the key challenge today is to "unearth the significant contribution Marx, Engels, and subsequent Marxists have made to ecological thought in the belief that a Marxist framework allows for the most coherent and useful modality for understanding the roots of the ecological crisis and plotting a way out of it."[27] In *Ecology and Socialism* that Williams observes that "ecological devastation is not an accidental outcome of capitalist development but an intrinsic element of the system, just as integral as class exploitation, poverty, racism, and war."[28] A socialist outlook brings to ecological politics specifically *global* and *dialectic* elements grounded in systemic analysis of world capitalism: no reversal of the crisis is possible within the existing order, ruling out prospects for a "green capitalism." For Marx and Engels, it was in the very nature of capitalism to expand the production apparatus, driven by profits, wealth, and power exercised not only over the economy but over government, social life, culture, and nature. Capitalism operates according to the logic of commodification, which opposes public initiatives that might impede this process. Global crisis is entirely predictable once the historical relationship between capitalism and the natural habitat, between private interests and social needs, is fully taken into account.

Marxism and the divergent socialist paths it inspired foresaw intensification of class struggle between workers and capitalists, a prelude to insurrection to overthrow the dominant order. At the same time, in its main political strategies (notably social democracy and Leninism) Marxism shared with liberalism an attachment to Enlightenment values—human progress through scientific rationality, technology, and industrial growth. It also embellished strong currents of the Judeo-Christian tradition, most centrally a drive toward mastery of nature. For classical Marxism, human alienation could only be abolished through overturning the capitalist division of labor, a necessary stage in the full realization of species-being, or liberation within classless society. Nineteenth-century socialists—not only Marx and Engels but Karl Kautsky, George Plekhanov, and others—inherited a modernizing faith in the power of science and technology, in the blessings of material prosperity historically generated by capitalism itself. The egalitarian side of Marxism signaled a radical shift in approaches to human community, but the theory sought to redefine human-nature relations, predictable given ideological constraints of the period. The positivist, scientific side of Marxism, wedded to Enlightenment rationality, militated against such reformulation. Marxism was resolutely productivist in its fixation on economic forces as the driving mechanism of history, central to the transition from capitalism to socialism.[29] Such an outlook meshed with the *Zeitgeist* of the period: Marxism, after all, gained ascendancy during the early modern period, forged between 1840 and 1880 and reached its zenith in the decades preceding World War I. The theory resonated with strong intellectual currents of the time and place (Europe), including a firm optimism in the liberating power of science and technology.

It has been argued that Marx (and later Marxists), despite ideological limitations of time and place, arrived at a conceptual framework universally relevant not only to class struggle and revolution but to ecological change. The socialization of production, a shift toward egalitarian class and power relations, breakdown of the division between urban and rural life, the overcoming of alienated labor—all this is said to point toward a model of sustainable development consistent with harmonious relations between society and nature.[30] Whether this imputed vision, usually based on very schematic and overly general passages in Marx, effectively counters a productivist obsession with economic growth is problematic, but even if we recognize an ecological Marx we are left with his (understandable) silence on the question of developmental sustainability. There is little in Marx (or later Marxists) to indicate serious theoretical reflection on this question—nor indeed has anyone made such a claim. As Ted Benton, otherwise sympathetic to Marx, observes, the overall thrust of the theory is to give humans a much *freer* hand in utilizing the natural world for human purposes, with class struggle in fact a vehicle for the "humanization of nature."[31] The much-celebrated "humanism" of the early Marx actually replicated a deep-seated ethos of human domination of nature found in Western religious and philosophical thought, including the Enlightenment. For Marx, following the legacy of Descartes, Kant, and Hegel, humans are innately creative and self-reflective, potentially free to make history, while nonhuman life remains trapped in a passive, predesigned biological realm. Instead of an organic connection between humans and natural life, Marx and Engels stressed dualism and *opposition* between the two—a tendency that would become more pronounced in later, more crudely deterministic, variants of Marxism.

True to the Enlightenment tradition, early Marxism associated prospects for social progress—including the historical struggle for socialism—with an expanding human capacity to master the natural world. Capitalism had already unleashed new material and technical powers allowing humans to transform and reappropriate nature. The labor process itself, a crucial site of technological innovation, contained within it the vast potential for human domination of nature, people gaining control over the environment (and themselves) in the process of achieving self-activity and liberation for, as Engels observed, under socialism humans would finally become the "true masters of nature." William Leiss writes that in classical Marxism the development of class struggle itself was seen as broadening human control over nature that, however, was destined to generate new contradictions within an expanding industrial system.[32]

A further, and in some ways more problematic, set of issues stems from the failure of early Marxism to articulate a political course forward, to provide a theory and strategy of change beyond general references to class struggle and revolution. It never arrived at a concrete framework establishing how the transition from capitalism to socialism was expected to unfold. Most centrally, the issue of class consciousness—vital source of revolutionary transformation—was scarcely posed, much less resolved in part owing to a psychological rationalism that *assumed* formation of an anticapitalist proletariat. How could ideological transcendence among workers (or any group) occur within parameters of a capitalist system exercising overwhelming material, institutional, and ideological domination? Neither Marx nor Engels fully addressed this problem, apparently convinced that an alienated proletariat would be naturally driven toward opposition and revolution, with no "external element" (vanguard party, intellectuals) theorized. The difficulty here is that history, then and later, revealed the inadequacy of this reductionism once the imperative of choosing political option came to the fore. By the end of the nineteenth century it was clear that workers, embedded in the daily rhythms of proletarian existence, would end up trapped in their social immediacy. The transition to socialism would require a definite *political* framework, build around effective organization, leadership, strategy, and methods for winning power. Without such a framework, oppositional forces would most likely be fragmented, neutralized, and depoliticized, however valiantly they struggled to improve their working and living conditions.

It would be remain for later Marxists to formulate distinct political strategies based on their understanding of historical conditions and social forces at work. Between 1890 and World War I rival strategic views emerged within Marxism (more accurately, socialism)—views subject to ideological debate and political conflict for decades after. Eduard Bernstein's evolutionary socialism, the theoretical genesis of social democratic reformism in Europe and beyond, resolved the dilemma of class consciousness by linking politics to immediate reform struggles—to the realm of elections, trade unions, and parliamentary work. Socialism would be achieved through a gradual and peaceful transition, based in democratization and socialization of liberal-democratic structures. As industrialization gave rise to an expanded working class, electoral victories and union gains meant anticapitalist victories and socialist gains, legitimated on a foundation of nationalist legitimation. Class consciousness would be taken as already constituted with the expectation it would evolve into a mature socialism as social democratic victories accrued. Similar to liberalism of the period, social democracy was drawn to the Enlightenment project of industrial growth through science and technology, to be realized within the nation-state system. Social democratic parties belonging to the Second International flourished across the twentieth century, the architects of a welfare-state capitalism that, in the end, fell short of Bernstein's vision of an evolutionary socialism.

The solution to the strategic dilemma that Lenin and the Bolsheviks preferred went to the other end of continuum: if workers did not exhibit revolutionary consciousness, as Bernstein argued and Lenin agreed, then the one truly *socialist* option would be to build a vanguard party charged with bringing such consciousness to workers otherwise trapped in bourgeois ideology. Party leaders would furnish ideological cohesion, organizational integrity, and a political strategy geared to the insurrectionary conquest of state power. Lenin believed that Bernstein's evolutionary schema was mere capitulation to the ruling interests, ensuring a lapse into liberal reformism while potential opposition was confined to the bourgeois public sphere. The capitalist state was not reformable from within its own rules, norms, and institutionalizing logic. Left to their spontaneous and economistic impulses, workers could never become part of a revolutionary process. The very idea of a peaceful transition to socialism was a dangerous myth, as the ruling class will mobilize everything at its command, including the military and police, to defend its wealth and power. For Lenin, the main conduit of revolutionary change was a combat party led by a Marxist intelligentsia deeply schooled in theory and strategy, fully prepared for organizational warfare—a recipe that initially succeeded, paving the way for a Third (Communist) International centered in the Soviet Union.

As for social democracy, the masses were at least theoretically anointed protagonists of history—their efforts mediated, to be sure, by parties and unions—but within a nonrevolutionary framework, whereas for Bolshevism these same masses were subordinate to a vanguard responsible for integrating and controlling political action. Jacobinism endowed the socialist intelligentsia with historical primacy, as popular forms (councils, unions, coops, etc.) were ultimately reduced to "transmission belts" managed by the party-state. Once in power, this new elite stratum would move toward overthrow and destruction of the old centers of power (aristocracy, monarchy, capitalists, etc.). This trajectory was followed by twentieth-century revolutions led by Communist parties—Russia in 1917 and later China, Yugoslavia, Vietnam, and Cuba. Whatever their Marxist beliefs and policies, these revolutions were eminently nationalist, victorious by means of popular armed insurrection against foreign imperialism and dedicated to the twin, interconnected goals of independence and modernization. These were simultaneously *multi-class* formations based not only in the urban working class and rural peasantry but in sectors of the middle strata, intelligentsia, and even national bourgeoisie. Communist revolutions were fueled by a Marxist-Leninist ideology that, with the party-state often under siege, retreated from more emancipatory (democratic, egalitarian, internationalist) goals. In the case of the Soviet model, this "great retreat" was set in motion already by the late 1920s when Stalin had consolidated his authoritarian rule. With industrialization, moreover, came the very human domination of nature that characterized both liberal-capitalism and social democracy. Ecological priorities received only token attention, which meant worsening habitat destruction, terrible air and water pollution, blighted cities, and, in the case of the USSR, nuclear disaster. Despite many differences, social democracy and Leninism eventually came to share crucial features, both vehicles of an Enlightenment that underpinned historic goals of bourgeois revolutions in France and America. Leon Trotsky was not the only radical to observe that both party leaderships had abandoned socialist internationalism along with the ideal of a genuinely classless society. A shared fetishism of technological rationality coincided with capitalist reliance on the very social hierarchy, organizational discipline, and bureaucratic efficiency that defined the bourgeois industrial order—more pronounced in Stalinist USSR than in Western social democracies where pluralism and citizen participation were further developed. Environmental devastation, whether in the USSR, China, Eastern Europe, or Western Europe, was the inevitable legacy of ostensibly socialist-driven modernization.

Twentieth-century Marxism gave rise to a third strategic path—syndicalism, or council communism—understood by its theorists as a mass-based, democratic alternative to the bureaucratic, centralized, statist direction of both social democracy and Leninism. Identified with such figures as Rosa Luxemburg, Georges Sorel, the early Antonio Gramsci, and council partisans like Anton Pannekoek, this path drew heavily on classical anarchism deemed consistent with early Marxism in its vision of a classless, stateless society. Rejecting the primacy of intellectual/vanguard elites, the council model affirmed faith in spontaneous popular activity, in contrast with the more jaundiced view of mass consciousness held by Bernstein and Lenin; party elites, no matter their ideology, inevitably worked against egalitarian, democratic principles needed for a socialist order. Revolutionary change would have to be prefigurative, anticipating the future in the present, to be consistent with its ideals. For Luxemburg, both reformism and Jacobinism betrayed mass revolutionary impulses and guaranteed the ascendancy of a new ruling elite that would exploit the very workers it claimed to represent. Council theory held that the masses would gradually be moved toward heightened class consciousness as mounting contradictions of a decaying capitalism gave rise to crisis, breakdown, and political opposition. Contrary to the assumptions of Bernstein and Lenin, workers in capitalist society were *not* condemned to a narrow economistic or bourgeois consciousness. Owing to its cherished spontaneism and fear of statism, this strategy embraced distinctly *local* forms in the transitional process—workers' councils, popular assemblies, and so forth—anticipating a new society based on self-managed networks of governance, with democratic citizenship informing every realm of public life.

The council tradition thrived in different regions of Europe after the turn of the century, reaching its peak in Russia before and during the October Revolution, in Italy during the explosive *Biennio Rosso* (1918–20), and in Spain during the Civil War (1936–39), but aside from these altogether brief (and ill-fated) moments of success it would eventually lose its political vitality. Given problems of extreme fragmentation, provincialism, and organizational fragility, the councils turned out to be impotent against more concentrated bastions of power. They were crushed by the state in Russia and Spain, defeated because of their own inertia and futility in Italy, and wound up assimilated into deeply entrenched union and party structures in Germany. Despite their promise of a democratic alternative, local forms alone have never provided the foundation of durable political organization or governance in any setting.[33] Moreover, while partially inspired by the ecologically friendly theories of Rousseau, utopian socialists, and social anarchists like Kropotkin, council communism itself generated no useful or lasting ideas pertinent to ecology or sustainable development. Despite historical failures, however, the council model in different guises would be rekindled decades later within the context of new-left radicalism and later new social movements.

Twentieth-century Marxists were no more likely to confront ecological challenges than were earlier theorists: "Western" Marxists like Luxemburg, Karl Korsch, Georg Lukacs, Gramsci, Jean-Paul Sartre, and Herbert Marcuse took up a wide range of nonproductivist concerns—popular culture, aesthetics, philosophy, the family, media, et cetera—but, with the partial exception of Marcuse, seemed hardly more interested in ecology than nineteenth-century Marxists, despite some forays into the discourse of "nature." There was no identifiable ecological sensibility, nor any effort to develop one—hardly astonishing insofar as environmental problems were obviously much less visible at the time. Western Marxism, meanwhile, reflected a deeper conundrum—acknowledgment that Marxist theory in general had failed to produce a distinct *revolutionary* politics for advanced capitalism. Once the failures of both social democracy and Leninism (and the later Soviet model) are recognized, the question of remaining strategic options begins to surface. In this theoretical vacuum it was critical Marxism, itself largely detached from political activity, that drew attention to deficiencies in the Marxist tradition *tout court:* weakness of socialist consciousness among workers, tendencies toward bureaucratic hierarchy and statism within organized Marxism, misplaced crisis theory, and the gradual solidification of bourgeois hegemony with capitalist industrialization. One result was a relative absence of mass-based revolutionary parties in the capitalist world. By the end of the twentieth century modern capitalism had been able to deflect crisis and integrate potential opposition, owing to its alignment with state power and its greater capacity for ideological control. It was the signal contribution of Western Marxists to identify and analyze this historical problem, each theorist arriving at different conceptual frameworks: Gramsci's "ideological hegemony," Lukacs' theory of "reification," the Frankfurt School's emphasis on "mass culture," and Marcuse's "technological rationality" (or "one dimensionality") being the most prominent.[34] Powerful as the Marxist legacy has been, therefore, the search for a broadly radical ecological politics appropriate to the later, far more destructive, phase of capitalism would have to proceed along newer intellectual and political tracks.

One result of this predicament was that "the environment" would be taken up by theorists (and activists) *outside* the Marxist tradition, since for Marxism (and the organized politics it spawned) social transformation was a project for and by humans struggling to conquer nature—"conquest" in the service of science, technology, and industrialization leading to whatever progressive social gains might follow. Nowhere, as mentioned, did issues related to the ecological crisis enter the political agenda: where addressed, it was assumed they would be "solved" through socialism, at a point when capitalist interests no longer prevailed. Communist leaders, for their part, viewed such issues as a bothersome distraction from more urgent (material, national) challenges at hand. By the time writers like Carson, Bookchin, and

Commoner began calling public attention to ecological problems in the 1960s, Marxism was already in serious decline, within and outside the Soviet orbit. Efforts to fuse ecology and Marxism, or socialism, would not be undertaken until later, as the underlying productivism and labor metaphysic of Marxist discourse had imposed limits on its capacity to critically theorize natural relations. To be efficacious in the new setting, Marxism would have to engage myriad issues raised by the new social movements and an emergent green politics.

It had become clear that Marxism, like the "social" and "deep" variants of ecological movements, could not by itself provide the basis of a revolution in natural relations. The new ecological radicalism would gain currency by confronting the destructive consequences of industrialism and establishing the parameters for sustainable development. Its expansive, at times utopian, ideals revolved around themes of local community, environmental balance, mutual aid, limits to growth, and generalized revolt against domination. Yet it also engages Marxism, notably in its critique of capitalism and class exploitation. What is often labeled "ecosocialism" passes through the seminal contributions not only of Marx and Engels but of Rousseau, utopian socialism, anarchism, feminism, and more recent ideas associated with modern anarchists, the counterculture, and social ecologists like Bookchin. It embellishes an antiauthoritarian politics, opposition to the entire matrix of domination, and the struggle for harmonious relations with nature—arriving at what might be called a "left-green" ecological outlook grounded in diverse movements for social change.[35]

My argument here is that a new theoretical synthesis is urgently needed if the modern crisis is to be fought with any hope of success—a synthesis integrating progressive elements of neo-Marxism, social ecology, and deep ecology infused with democratic sensibilities. World capitalism has grown more corporate-driven, authoritarian, violent, and unsustainable in recent decades, much of it centered in the United States with its great concentration of transnational economic power, its vast banking centers, its bloated warfare state, and its sprawling corporate media system. This reality ought to force political strategy toward radical alternatives informed by a multiplicity of traditions and movements. We have entered a capitalist phase that is increasingly chaotic, conflict ridden, and destructive. Immanuel Wallerstein observes, "The primary characteristic of a structured crisis is chaos. Chaos is not a situation of totally random happenings. It is a situation of rapid and constant fluctuations in all the parameters of the historical system. This includes not only the world economy, the interstate system, and cultural-ideological currents, but the availability of life resources, climatic conditions, and pandemics."[36] Wallerstein goes on to add that systemic crisis is likely to "increases the viability of agency."[37] Yet "agency," to be politically decisive, requires ideological articulation, organizational cohesion, and strategic direction. An ecosocialist alternative fits these criteria, but in its existing modalities lacks such coherence except where it becomes part of the Green orbit. Insofar as the crisis is rooted in multiple and overlapping conditions, opposition means moving along diverse fronts: politics, the economy, global relations, culture, the media, the environment. Ignoring this totality guarantees one form or another of political futility. Against the most extreme deep ecological currents, however, it seems clear that transformative change can occur only within the parameters of the existing urban, modernized order, rather than as a simple "exit" from that deeply entrenched system.

Whatever its theoretical limits or twentieth-century *political* fate, Marxism remains crucial to the modern ecological project, its class analysis and anticapitalist legacy vital to building anti-system movements against corporate power. The most vexing problems of the current period, including worker exploitation, global poverty, bureaucratic domination, wars, and ecological collapse, cannot be grasped in isolation, nor overcome in the absence of class-based movements challenging the hardened reality of corporate globalization. As John Bellamy Foster notes, "Socialism has always been understood as a society reversing the relations of exploitation of capitalism and removing the manifold social evils to which these relations have given rise. This requires the abolition of private property in the means of production, a high degree of equality in all things, replacement of the blind forces of the market by planning of the associated producers in accordance with genuine social needs, and the elimination to whatever extent possible of invidious distinctions associated with the division of town and country, mental and manual labor, race relations, gender divisions, etc." Much of this is rather explicit in Marx and especially later Marxists. Foster adds, "The only way to accomplish this is by altering our human metabolism with nature, along with our human-social relations, transcending both the alienation of nature and of humanity."[38] This addendum, however, is at best only vaguely implicit in the writings of Marx and Engels, or indeed any earlier Marxists, while the organized political forms of twentieth-century socialism never articulated ecological concerns in either theory or practice.

Expanded corporate power has created the global conditions of intensified human alienation from nature symptomatic of deepening crisis. Foster, Williams, and others who might be labeled "ecosocialists" are correct to emphasize the urgency of a global ecological revolution dependent on full-scale transformation of class and power relations. Sustainable development is inconceivable within the existing world system, but such an eventuality, as Williams argues, "must encompass social sustainability, equality, and justice as much as it does ecological concerns," adding, "We can only do this if we collectively and democratically make all decisions based on human need not corporate profits."[39] Foster continues along these lines: "It is the historic need to combat the absolute destructiveness of the system of capital at this stage—replacing it, as Marx envisioned, with a society of substantive equality and ecological sustainability—which, I am convinced, constitute the essential meaning of revolution in our time" (p. 14). Such a process is imperative, but it is noteworthy that the ecosocialism embraced by Foster and Williams remains largely silent on matters of political strategy, including what is to be the agency (or agencies) of ecological revolution. What methods of struggle and approaches to power are needed to face mounting challenges when time is very short? General theoretical propositions ultimately demand concrete organizational and political translation. Ecosocialism might well pose the ultimate hope for humanity, but Foster, Williams, and others sharing this outlook have been frustratingly vague about strategic directions. Given the historical deficiencies of social democracy, Leninism, and council communism as modern-day political options, fresh thinking is imperative at a time when the planet nears the ecological tipping point. To date the main political expression of ecosocialism has been on the fringes of European Green parties—a tendency more fully explored in the next section.

Some deeper flaws in Marxism have been identified by theorists of both social ecology and deep ecology, with their more systemic approach to the environment, their eclectic concerns, and their sharpened attention to *multiple* forms of domination in modern society. Sensitive to the complex *ensemble* of relations, social ecology resists the productivism and class reductionism that undermines prospects for a fuller ecological Marxism. Meanwhile, deep ecology advances a more comprehensive view of natural relations, extending normative status to wide areas of nonhuman nature. Neither Marxism nor social ecology rival DE in the sense of gravity it attaches to habitat destruction, in its deeper critique of the Enlightenment and industrialism, the perils of technological fetishism, and the obsessive pursuit of material abundance. Only DE, moreover, is uncompromising on limits to growth, calling for alternative modes of agriculture, production, and consumption in harmony with the earth's carrying capacity—a viewpoint never clearly formulated within the Marxist tradition. Such a qualitative shift in social, ecological, and indeed *political* arrangements can be met only with something akin to a post-Marxist radicalism.

A Green Synthesis?

With the ascent of modern environmentalism in the 1960s, it seemed only a matter of time before the birth of a political party, which soon enough materialized in the West German Greens during the early 1980s. The route to organized politics was in fact surprisingly quick, emerging as it did within a society governed by an institutionalized power structure ruled equally by the Christian Democrats and Social Democrats. The Green electoral breakthrough in 1983 helped legitimate a party that was the first to place ecological agendas at the center of its political ideology. Building on a vibrant gathering of oppositional movements-environmental, peace, feminist, community groups—the Greens departed from both social democracy and Soviet-style Communism on the left as well as from the German corporate state, labeled "Modell Deutschland." Referred to as something of an "antiparty party" at its origins, the new formation upheld hopes for an ecological radicalism fueled by commitment to sustainable development and vision of democratization extending beyond the realm of parties and elections, grounded in local movements. This was a novel synthesis of electoral politics and popular struggles, meaning a fundamental break with the corporatist state. Winning a series of local and federal contests while building a number of "red-green" alliances with the SPD, by the end of the 1980s the West German Greens had conquered new political terrain and, just as important, had raised questions about the nature of power, citizenship, and social progress in the context of natural relations. Their breakthrough soon extended to other nations in Europe and beyond, posing a crucial question: could Green parties sustain their presence as a radical force beyond the initial euphoria of electoral victories?

In March 1983 the West German Greens won enough votes (5.6 percent) to enter national parliament (Bundestag) and give institutional voice to grassroots movements that had begun to flourish in the 1970s. The party could now boast two million supporters, 30,000 members, and 27 Bundestag deputies. By 1985 they managed to build on this success, winning new footholds in state and municipal legislatures, local movements, and the general political culture, despite no corporate resources or favorable media coverage. Green leaders like Rudolf Bahro, Daniel Cohn-Bendit, and Petra Kelly unfurled a strong critique of the SPD, which by the 1980s had degenerated into another catch-all party with no socialist identity, while also ruling out a Leninist-style insurrectionary seizure of state power. What the party shared with most grassroots movements was an emphasis on nonviolent direct action, a struggle for local empowerment against rampant urbanization, opposition to corporate-driven modernization and consumerism, hostility to interest group bargaining, and skepticism of conventional ideologies including liberalism, socialism, and nationalism. Any political scheme that might be associated with bureaucratic power was summarily rejected.

Once the Greens reached a position from which they could pursue a dual strategy organically linking party and movements, elections and grassroots struggles, legislative reforms and direct action protest, internal divisions inevitably surfaced. Splits fell along different lines—between ecological "Fundis" dedicated to a holistic politics, ecosocialists emphasizing class and material issues, and pragmatic "Realos" in favor of tangible reforms; between those seeking closer ties with the SPD and those hoping to preserve Green autonomy and identity; between those looking to parliament and those immersed in the social movements. What served to unify the majority of activists and supporters, however, was an abiding dedication to ecological politics. Further, the Greens' post-liberal and post-Marxist understanding of multiple and overlapping forms of domination clashed with both interest group pluralism and the single dialectic of class struggle associated with Marxism. Bahro had said, "I think it has become very doubtful that the proletariat within bourgeois society will be the bearer of the subject of a new society," adding, "The class struggle is not the solution."[40]

At each level—federal, state, municipal, communal—the Greens set out to bring representative democracy closer to the grassroots, opening up debates, sharing material resources, building local committees, and looking to rotate leadership with a view toward gender equality. Parliament would be an arena for mobilizing general ideological consensus around issues as much as for winning power. Electoral successes conferred on the Greens a credibility that could be used, though not without pitfalls, to empower and politicize local citizens' initiatives. Such an extension of democratic politics was scarcely, if ever, advanced within either Communist or social democratic strategies. This early Green outlook probably owes more to new-left radicalism (and its sequel, the new social movements) than to any other ideological current, detectible in several areas: the motif of overcoming human alienation, the ideal of "dual power" or counter-society, emphasis on deeper forms of citizenship, an open and eclectic political style, attention to personal politics, and principled dedication to nonviolence in all realms of human activity. This last principle was a cornerstone of Green strategy insofar as violence or coercion, even in the service of change, was thought to replicate the culture of domination that pervaded existing power relations; it would only beget more violence and authoritarianism. Further, while transition to a new society would obviously involve disruption, conflict, and mass mobilization, armed insurrection was regarded as counterproductive.

Integral to this transition was an effort to formulate something along lines of an ecological model of development. The global crisis called for a shift from a corporate-growth economy rooted in consumerism and arms production to one more rationally designed to meet human needs and environmental sustainability. For the Greens, a capitalist model orchestrated by the large enterprises, banks, and the state was a primary source of crisis, a system taking the planet beyond the ecological tipping point. Economic proposals favored increasingly socialized, decentralized arrangements, expanded public allocations, reduced growth, and forms of grassroots self-management, though specific measures were often lacking. The Greens looked to a process of conversion, with material resources directed increasingly toward basic needs such as health care, housing, transportation, education, and public infrastructure. Slower growth would thus allow for enhanced quality of life along with a healthier natural habitat. But here, as in other areas of Green politics, methods for achieving such objectives were only vaguely set forth. In one area, however—alternative technology—the party not only laid out clear policies but took on a leading role: renewable energy sources (wind, solar, etc.) would be the basis of a "democratically controllable" system in which "ecological accounting" would replace standard fiscal measures.

If European Green politics involved a merging of disparate social forces and ideological tendencies, its initial gains were fueled by the huge peace mobilizations that swept Europe during the early 1980s. Protests against NATO's planned nuclear deployments helped galvanize the Greens, in West Germany and across Europe. There was a growing alarm over escalation of the arms race, the nuclear threat, renewed Cold War tensions, and resurgent U.S. militarism under President Reagan. While issues connected to the U.S.-Soviet rivalry naturally vanished in due course, the question of military spending and arms production would persist. The Greens argued that each new cycle in arms production heightened contradictions between economic dysfunction and social progress, revealing a systemic linkage of corporate interests, militarism, and global crisis. No other political entity, in Europe or elsewhere, had called attention to this linkage. It was Bahro who predicted that "the peace movement is now at the head of an entire social constellation that is rehearsing the emergence of a new epoch."[41]

Moving from these premises, the Greens called for a bloc-free Europe coinciding with a "nuclear-free zone stretching from Portugal to Poland," the cessation of arms exports and nuclear reactors, and initiation of worldwide disarmament talks. Their peace overtures were intended to destroy the feeble rationale for NATO, reverse the arms race, and (ultimately) dismantle the military apparatus one country at a time. The long-term goal was a nonmilitarist defense policy rooted in the concept of social defense relying on disarmament, social conversion, nonviolence, and the abolition of military blocs. The essence of social defense, as elaborated by Gene Sharp and others, was formation of social networks to mobilize broad civilian resistance against military incursion from outside.[42] Petra Kelly among others insisted on an unwavering commitment to nonviolence in the Gandhian tradition—not merely as political method but as general "defense" policy, a moral vision, and way of life. Where opponents of a violent order choose violent methods they destroy their own legitimacy, for politics, as Kelly insisted, must involve unity of ends and means in a way that prefigures the future.[43]

Green entry into the established political system provided new ideological space where long-repressed issues could be raised and debated, the power structure could be more directly confronted and details about its operation more widely disseminated. As party membership reached 42,000, the Greens received more than three million votes in the 1987 federal elections and would soon have 7,000 elected officials at all levels. There were local "red-green" governing coalitions, leading to national collaboration between the SPD and Greens. In most locales the Greens became the third strongest political organization. Meanwhile, this new model—given momentum by the ecological crisis—quickly spread across Europe, to Sweden, Finland, Holland, Belgium, Austria, Italy, France, Ireland, and Greece, with parliamentary representation in most cases achieved. Rival leftist parties (notably Socialists and Communists) made new efforts to bring ecological concerns into their election platforms. Meanwhile, a series of dilemmas soon accompanied Green organizational and electoral growth. How could the parties sustain grassroots mobilization while being so deeply involved in parliamentarism? What strategy would they adopt toward other leftist groups, trade unions, and the labor movement? How could a radical ecological identity be preserved in the face of deep conflicts around pacifism, spiritualism, and grassroots democracy? The capacity of Green leaders to creatively deal with such challenges would determine the party's trajectory—and the immediate results would not be especially hopeful.

The majority Green outlook, in Germany as elsewhere, favored some type of red-green alliance that might allow the nascent parties to avoid isolation and impotence, especially where partnerships were viable and the Greens could enter into governing coalitions. The Realo position was more disposed to compromise, partnerships, and casting broad electoral appeals to bolster Green expansion. The Fundi wing, on the other hand, feared that collaboration with the SPD would stifle Green identity insofar as the party needed space to advance its unique style, interests, and goals. Bahro, among others, believed it was the historic task of the Greens to *replace* the social democrats as Germany's main oppositional force, so that cooperation would undermine this imperative. Fundis were convinced that, whereas the SPD looked exclusively to the state, there could be no merging of interests with a Green Party still embedded in local movements. Further, while the SPD's own corporatism and statism made its leaders fearful of change—of any policies that might offend capital—the Greens were dedicated to qualitative (ecological, feminist, grassroots) demands aligned with the emergent citizens' struggles. The Fundis were no doubt correct to observe that long-term Green participation in electoral and parliamentary activity would likely bring deradicalization—the very fate of postwar European Socialist and Communist parties (including the SPD). There was little reason to believe the Greens would be any more immune to the pressures of institutionalization and moderation than were these other parties.

In nearly three decades since the Green breakthrough, such deradicalization has indeed been the trajectory of parties in Germany and elsewhere, whatever their influence on public policy and the political culture. With a steady electoral presence at the municipal, regional, and national levels, the Realo faction managed to establish broad control of party organization, reinforced by a series of red-green coalitions that produced a federal Green-SPD governing alliance from 1998 to 2005. The Greens had become fully integrated into the parliamentary system, both locally and nationally, as their ambivalence toward electoral politics waned. The outcome was predictable enough—a more conservative organizational style, ideological moderation, detachment of party leadership from the mass base, professionalization of the entire structure. The very originality of Green politics, which spoke a language of "break" or "rupture" with a corporate-state past, had by now vanished. The idea of a distinctly post-liberal, post-Marxist ecological politics, holding so much promise in the early 1980s, became a distant memory. Not only the domesticating logic of electoral politics, but the gradual decline of new social movements, contributed to this atrophy. Viewed thusly, by the early twenty-first century the Greens in Germany and beyond had turned into replicas of the very social democracy they initially so despised.

After the early 1980s, as mentioned, the Green phenomenon has assumed a global character, spreading to numerous countries around the world—though not according to any single party model. The first European Congress of Greens, held in April 1994, attracted delegates from eight European countries looking for broad organizational (though not necessarily programmatic) unity. The very eclectic and amorphous character of Green ideology defies any unitary pattern or formula typical of the Second and Third Internationals. A European Green Party was formed in 2004, comprising more than 40 parties across the continent. After winning 26 seats to the European Parliament in 1989 and then 38 seats in 1999, the Greens were able to elect 46 deputies in 2009. They benefited from a decided upturn in popularity since the late 1990s owing to the worsening ecological crisis, but party ideology has been diluted since the early years. Grassroots dynamism from the peace, environmental, feminist, and community-based movements has visibly weakened. Previously opposed to European integration (as a capitalist stratagem), the German and most other Green parties have come to accept it. Once resolutely against NATO military deployments and operations, many Green leaders endorsed the 1990s U.S./NATO interventions in the Balkans and Afghanistan. The motif of grassroots democracy has been upheld in theory but mostly forgotten in practice.[44]

In the United States, a Green alternative took shape in 1984 at the historic gathering of the Greens Committee of Correspondence. There was receptivity to the idea of a "new politics" growing out of mostly new-left concerns, popular movements, and the counterculture—along with the American traditions of civic participation, local community activism, and self-help. At the same time, an emergent U.S. Green Party was operating in a relatively conservative milieu long hostile to anything resembling European socialism. Adopting the slogan "neither left nor right," American Greens looked to different modalities of new-age spiritualism while shying away from the anticapitalist critique accepted by most Europeans. There would be no pressures toward a red-green alliance in the United States, where social democracy (like Communism) has a weak history, but uniform reliance on electoral politics gave rise to incremental local gains as hundreds of Green candidates were voted onto city councils across the country. If the American party had not yet achieved radical oppositional status, neither did it move very far beyond the liberal environmentalism critically explored in Chapter 3. Their idea of "sustainability" would be entirely compatible with established political and economic arrangements, requiring no transformation of the corporate-state system.

The emergence of the Occupy movement in late 2011 has been something of a stimulus to Green politics in the United States. For one thing, the famous "Green values" systematically coincide with the ideological tenets brought forward by the Occupy activists, including ecological wisdom, decentralization of economic and political power, respect for diversity, and nonviolent resistance. While the long-term consequences of this movement for organized Green politics remains unclear, evidence show that a large number of Green partisans have become involved in the Occupy insurgencies as they have spread across the United States. With the appearance of two distinct (and often competitive) Green parties, it is the original formation organized mainly around local organizing—the Greens/Green Party USA—that resonates most closely with the Occupy tendency. This formation has been especially hostile to the idea of mounting high-level, national electoral campaigns of the sort run by Ralph Nader in 2000. Nader's campaign was sponsored by Green Party of the United States, which broke off from the original Greens in 1996 in preparation for Nader's campaigns that year and in 2000—the latter blamed by many for George W. Bush's victory. In this case, as in many others, the wrath of establishment politics and media came down generally on Greens politics, once again reflecting the enormous burdens and roadblocks faced by anticorporate third parties in the United States.

Meanwhile, European Greens managed to keep alive ideals of ecological radicalism that originally brought several parties to electoral prominence. It is worth emphasizing that Green politics, even in

Europe, is still rather embryonic, with international collaboration even more recent; most parties came into prominence long after the initial German successes. Today European Greens continue to uphold a program of "eco-development": environmental sustainability and limits to growth, alternative technologies, social justice, local economic forms, grassroots democracy. They urge moves toward "global security" including an end to military blocs, nuclear disarmament, a full-scale test ban, and stepped-up efforts to prevent armed conflicts. As before, they celebrate a deeper citizenship rooted in community activism and popular movements. If this sounds radical enough, it is tempered now as before by strategic vagueness, with basic questions about economic and political power commonly sidestepped. To the degree that an institutionalized state-corporate order occupies the core of the global crisis, it will have to be confronted more directly—perhaps even more violently—than what most European Greens seem prepared to undertake. Green politics as it remains today will need more clearly articulated strategic content for it to be historically viable.

Envisioned as a novel party based that merges electoral politics with social movements—a post-liberal, post-Marxist opposition—the Greens have for three decades embodied the closest thing the world has seen to a mature, strategically defined ecological radicalism. Despite limits and flaws, they seem to constitute the only political force, with some global presence, dedicated to reversing the modern crisis—and the only force with a coherent strategy for change. Their electoral gains, however, have been mitigated by ideological erosion as radical identity is difficult to sustain, in Germany and elsewhere. If a transformative Green politics is to take root and flourish, it will need a sharper anticapitalist outlook and a jettisoning of less-helpful discourses such as the vacuous slogan "neither right nor left." Success could well depend on a convergence of leftist (social-ecological, democratic-socialist) and deep ecological currents, as well as a merger (only episodically witnessed to date) of progressive elements in diverse social movements: labor, antiwar, Civil Rights, feminist, community action. Too many Green leaders have been obsessed with purging "red" tendencies linked to the Marxist tradition that, as I have argued, offers indispensable class analysis and global perspective.

Green successes in Europe and elsewhere, blessed with more open electoral and parliamentary systems, is one thing—prospective gains in the United States, with its tightly integrated two-party system, yet another. Even the vast majority of American progressive despair at the thought of breaking through the vaunted party duopoly of Republicans and Democrats, both equally ruled by corporate interests. There is the vexing question of precisely how a Green alternative might hope to achieve enough political leverage in the United States to shape an ecological agenda. The duopoly, after all, is set up to push electioneering toward the mythical "center": it moderates, colludes, neutralizes, delays, and above all resists change—exactly as the Founders, ever fearful of mass insurgencies, had imagined. If the two-party oligarchy is designed to negate third-party enterprises, then hopes for a Green synthesis gaining real power would appear dismal. In fact the history of third-party efforts—Progressives, Socialists, Communists, Libertarians, Greens—has been one of uniform historical failure, and few such alternatives have lasted more than a few election cycles, and none has managed to gain much of a foothold in the political system. The great historical successes of liberal capitalism have given rise to a powerful ideological consensus not only at the summit but across the political culture, helping to explain the virtually all-encompassing hegemony of corporate and financial institutions in the United States today. Such ideological factors converge with structural arrangements—single-member voting districts, a Presidential system, skewed representation in favor of conservative interests, et cetera—that render third-party access nearly impossible. It follows that any break with this system, allowing for a Green breakthrough like that in Europe, would require first and foremost a social crisis of nearly unprecedented proportions that would force critical masses of people away from politics-as-usual. A congruence of mounting crises today (economic, military, political, environmental) might conceivably be the catalyst of such a break. At this writing (late 2011) voting surveys indicated that party identification for both Democrats and Republicans has been slipping gradually while the percentage of self-defined "independents" (38 percent) continues to rise, suggesting potential wider interest in challenging the duopoly. Still, efforts by Greens and others to carry out a serious challenge to the established oligarchy will likely fall short of historical success until crucial *structural* obstacles are removed, allowing (minimally) for a system of proportional representation that enables smaller vote-getting parties to elect candidates and build from there.

For an ecological politics vibrant enough to galvanize mass insurgency in the face of impending global disaster, the symbols of "red" and "green" do not appear as hostile opposition but as complementary ideological strains within a broader radical tradition, assuming that by "red" we refer to an outlook that is no longer strictly workerist, vanguardist, or social democratic. Deep ecological sensibilities will be central to any future Green synthesis, but—in contrast to its more utopian discourses—there can be no simple "exit" from modern industrial society, only the potential to socialize, democratize, and ecologically restructure the world system. This calls forth a political strategy close to that initially developed by the West German Greens, who set out to create an alternative to barbarism in realistic yet visionary terms. Despite myriad obstacles and detours, they did contribute to the emergence of a renovated public sphere in which transformative politics could be more openly theorized and practiced.

CHAPTER 6

A GLOBAL ECOLOGICAL REVOLUTION?

The journey toward an ecological radicalism will bear little fruit until it is translated into a modern, strategic—ultimately worldwide—form of *politics*. Revisiting the classic injunction "socialism or barbarism," the global crisis poses the question of human survival in a world order that is unraveling much faster than all but a few seem prepared to recognize. The environmental challenge has inspired, even forced, new ways of viewing not only economic development but political governance, culture, nature, and social change. A deeply ecological outlook invites radical perspectives on the future of production, consumption, agriculture, and technology, raising new questions about modernity itself as a product of Enlightenment rationality grounded in the utopian promises of science, technology, and material growth. The global crisis reveals the extent to which the classical industrial model has run its course, even as ruling elites scramble to mobilize resources in support of the corporate-growth system over which they preside—a system giving rise to rampant material exploitation, vast inequalities of wealth and power, wasteful use of natural resources, militarism, and warfare not to mention escalating habitat destruction on the road to possible ecological collapse. As Joel Kovel writes, "...the current stage of history can be characterized by structural forces that systematically degrade and finally exceed the buffering capacity of nature with respect to human production, thereby setting into motion an unpredictable yet interacting and expanding set of ecosystemic breakdowns."[1] There can be no exit from this predicament, no solution to impending disaster, without a departure from the past—from inherited and stultifying forms of economic development and political strategy.

As I argue in previous chapters, the familiar political legacies offer little basis for optimism: even "successful" revolutions like those under the aegis of Jacobinism failed to overturn class and power hierarchies. In the end, both social democracy and Communism reinforced and legitimated a dysfunctional industrial system that recognizes few limits to its institutional and material power. Mass-based alternatives—anarchism, council radicalism, the new left—were typically reduced to political futility. Green parties, with their promise of a two-pronged strategy linking party and movements, have so far followed the trajectory of social-democratic integration.[2] As for Marxism in its richly diverse theoretical expressions, any future revolutionary change will have to dispense with the old formulas (single agency of change, productivism, proletarian revolution) integral to the tradition. The era of organized state capitalism differs qualitatively from the classical model insofar as it brings to the fore intensifying *ecological* contradictions within the globalized economy. A mulitiplicity of social movements and ideological currents shapes the "postmodern" milieu, undermining political formulas based on single theoretical structures or political vanguards. Hopes for a global ecological revolution depend on the mobilization of wide support around plural identities, goals, interests, and appeals, precisely what the European Greens set out to achieve in the early 1980s. Today, as the ecological predicament reshapes the world landscape, humanity faces the imperative of radical change fueled by dramatic changes in economic, social, and political life. This in turn requires a developed *political strategy* that to date has attained only partial, uneven, localized expression.

The Recovery of Politics

In contrast to established political parties, social movements usually thrive on diverse styles, methods, and rhythms—often disruptive, episodically violent, and typically suspicious of state institutions. Movements can be highly unpredictable, scattered in their trajectory, and difficult to analyze. Some are counter-hegemonic in their *modus operandi*, subversive of prevailing discourses, institutional patterns, and social norms. However rebellious or moderate, oppositional or system sustaining, history shows that popular movements follow diverse paths, depending on political circumstances, social conditions, and tactical decisions. Movements as such have limited transformative potential, regardless of scope, definition, and levels of militancy, confined as they frequently are to their own social immediacy. It remains for more well-articulated political formations with cohesive organization, ideology, leadership, and strategy to bring collective efficacy to otherwise partial, spontaneous, often localized, struggles.[3] We have seen how the classic traditions (liberalism, social democracy, Leninism) fall well short of the political capabilities needed to reverse the modern crisis. Among other problems, they share an Enlightenment faith in the industrial-growth model and its continuous material expansion as the engine of human progress. Movement-based "alternatives," on the other hand—for example anarchism and council communism—never managed to escape a pre-political impasse, impotent in the face of large-scale state, corporate, and military power. It became the legacy of European Green parties, in theory post-liberal and post-Marxist, to forge a political synthesis compatible with radical possibilities that, however, were soon compromised by the logic of deradicalization. Green attempts to merge party and movements, elections and direct action, national and community struggles eventually faded under the relentless pressures of ideological and organizational accommodation. Such failures, however, hardly mean that political lessons cannot be gleaned from the historical experience.

If social movements ultimately require political translation to be historically efficacious, their very logic often resists such translation given strong elements of localism and spontaneism, marked by efforts to retain a precarious identity and autonomy. In the end, however, state institutions provide needed political leverage in the service of change. Strategies vary in their approach to state power, but movements *ignore* that realm at the cost of political impotence. In modern societies, as we have seen, the conventional "overthrow" scenario of twentieth-century Communist revolutions cannot work—or, if it did somehow "work," would give birth to an authoritarian behemoth much like the Soviet party-state. Given the social complexity of advanced capitalism, along with the democratic sensibilities of modern citizenry, no strategy for change could hope to smash state power—or simply ignore it: the only viable approach is to democratize institutions, processes, and norms. If state governance were imposed vanguard-style on civil society, hopes for a post-liberal, post-Marxist strategy would quickly turn to dust. The first European Green parties held to a local-democratic ethos but quickly fell victim to deradicalizing pressures, as the Realo wing gained ascendancy and reformist compromise prevailed over radical opposition.

Turning to American politics, the long and often dramatic legacy of social movements has been one of decidedly mixed outcomes.

Many grassroots struggles—Civil Rights, feminist, antiwar, environmentalist, community based—have reshaped the landscape enough to strongly impact policies, laws, and everyday public discourse. Such impact, however, has been typically more liberal-reformist than radical, nowhere more so than on environmental struggles. The search for leverage within the dominant public sphere usually corrodes antisystem energies as reformist gains come with stiff political costs. New-left insurgencies of the 1960s were yet another matter: student activism, the counterculture, and years of antiwar mobilizations were always pre-political in their well-known glorification of spontaneity, cultural radicalism, and deep hostility to government. With its unyielding antiauthoritarianism, the American new left worshipped democratic localism inherited from anarchism, the council tradition, and the progressive elements of liberalism, assimilating both strengths and weaknesses. And the weaknesses were abundantly visible in the triumphantly spontaneist character of movements, their rebellious fervor achieving little organizational and ideological durability, all too often limited by the very chaos and dispersion they celebrated.[4] Some disillusioned activists eventually chose Marxist-Leninist sects (Weathermen, Progressive Labor, etc.), but these withered from their own dogmatism, isolation, and irrelevance. By the 1970s, with proliferation of the "new movements" (notably feminism and ecology), the fetishism of extra-institutional struggles gave way to more developed political formulas including "pre-parties" like the New American Movement. It was precisely these "newer" movements that fueled radical politics during the 1970s and 1980s, including the European extra-parliamentary opposition and, eventually, the West German Greens.

New social movements remained vibrant through later decades, helping frame the dispersed "postmodern" social and intellectual landscape while detached to varying degrees from the realm of established politics—except where the Greens were able to organize citizens' initiatives into a loosely structured alternative party, an "antiparty party." Some Green parties were able to build what Gramsci had called a "social bloc" in which broad collective unity is forged out of turbulent social diversity.[5] In Europe the Greens entered government at all levels, sometimes as part of a "red-green" coalition aligned with social democrats and other leftist parties. In the United States, however, with its tightly integrated two-party system and greater solidification of the corporate-state, the smaller Greens were isolated from the political arena except for a scattered presence on city councils and similar local bodies. The Greens' dilemma in the United States was sharpened by a more pervasive corporate colonization of public life and the heightened role of electoral activity in legitimating elite power. Paradoxically, while new movements were considerably stronger in the United States than in Europe, their *political* articulation in the United States was always more feeble, more lacking in strategic efficacy. This was true even as the American Greens began to run candidates for Congress and the Presidency, clearly with no hope of winning.

Regardless of historical context, social movements (like political parties) have always faced the twin pressures of isolation versus assimilation, impotence versus deradicalization. If ecological radicalism points toward an overcoming of this dualism in the face of urgent global challenges, its capacity to resolve the dilemma of political strategy was not thereby made any easier. It was in fact rendered more difficult in the case of ecological struggles, where the incessant pull toward pure "nature" or "the country" and away from contaminating "politics" was often irresistible, as shown by the persistent appeals of a localist counterculture, deep ecology, new-age spiritualism, and "back-to-the-country" lifestyles.

Movements are inherently distinct from parties, but the former requires the latter in some modality to achieve historical efficacy. Movements can of course merge with parties, as in the case of the Greens, they can continue as pre-political formations (as local enclaves or interest groups), or they can simply vanish from the public landscape. Future radical change will be optimally served by the first alternative—a merger of party and movements, electoral and grassroots activity—which the European Greens first systematically introduced onto the scene. One-dimensional approaches are doomed to failure, incapable of transformative politics. In any event, recent American experience offers little cause for optimism. The three most-celebrated popular movements of the post–Cold War era—opposition to corporate globalization, antiwar protests, and immigrants-rights mobilizations—all sooner or later suffered from arrested development, falling dramatically (indeed fatally) short of political articulation. Consistent with the limits of postmodern ideology, these movements—mass based, diverse, and militant—followed a path best characterized as episodic, dispersed, and short lived, never achieving durable structure, cohesive ideology, political strategy, and focus on governance. In fact, all this was roundly dismissed by a critical mass of activists understandably resistant to the messy and corrupt domain of politics. Predictably, the movements wound up confined to pre-political (or, more accurately, antipolitical) futility, with little sustained presence beyond scattered local enclaves and interest group lobbying. After the explosive Seattle events, the so-called anti-globalization protests—euphoric but rather conflicted about goals and methods—essentially vanished from the American scene, briefly resurfacing as short-lived movements and forums at international business gatherings, as in Genoa during 2002.[6]

As for the high-energy peace mobilizations, their general rise-and-fall cycle is rather well known: going back a century or more, antiwar protests in the United States have typically fixated on the all-consuming war of the moment, driven by an activism that was ultimately ephemeral, dispersed, and futile. They surfaced within a largely negative and *reactive* framework—opposed to specific military interventions, opposed to nuclear weapons, opposed to missile deployments or weapons systems, and so forth. Deep economic and institutional conditions underlying the warfare state remained off the radar, with the actual causes of militarism and intervention rarely analyzed or confronted. The cycle of protest was destined to be repeated, ensuring that future wars would be the target of similar mass outpourings. As one war drew to a close, others could be anticipated. Little political activity has been directed against the war economy and the security-state, against the firmaments of U.S. militarism and warfare. Seymour Melman writes, "A peace movement in such a society can no longer limit itself to criticism of direct military combat. For this is no isolated condition but rather the continuous consequence of the dominant role of the war-making institutions in public life. Under these conditions 'peace' must, centrally, mean the diminished decision-power of the war-making institutions. That formulation, in turn, will help to overcome part of the ideological support system of the warfare state."[7] Melman long ago called for a distinct antiwar *political strategy* that, given an American culture attuned to immediate responses, local protests, and quick results seemed then (and now) completely out of reach. Looking at the spirited antiwar mobilizations against the Iraq war in early 2003, this fatal void was again dramatically revealed; millions of people took to the streets in several major cities, only to be followed by months, indeed years, of political quietism at a time when Washington could with impunity continue its military interventions in Iraq, Afghanistan, Pakistan, and Libya. College campuses, the epicenter of sixties anti–Vietnam War protests, marches, and teach-ins, had now become sites of overwhelming passivity and indifference.

The fate of huge immigrants-rights mobilizations against harsh state and federal legislative measures was scarcely more inspiring. Though large and militant, they were strictly *defensive*, bereft of any change-oriented vision or strategy. The movement lacked structural cohesion and durability: marches and demonstrations gave way to popular retreat, a return to business-as-usual, aside from the ongoing work of a few community groups.

At the final stages of this writing (January 2012), a series of new-populist mobilizations had burst upon the American scene, begun as Occupy Wall Street in New York and quickly spreading across the country, targeting the concentrations of wealth and power and quite possibly inspired by the Arab Spring. Tens of thousands of protesters, drawing on a long tradition of American challenges to the bastions of plutocracy, created a "leaderless movement" identified as the "99 Percenters" disgusted with Wall Street (and general corporate) fraud, theft, and corruption—harkening back to the Populist insurgencies of the 1890s and later, forced outside a two-party system beholden to these centers of power. From all appearances, the new populism is socially diverse, broad based, and dedicated, though highly diffuse and without much organizational cohesion and strategic direction. The "99 Percent Declaration" emphasized a wide range of urgent agendas, including a ban on corporate money in politics, medicare for all, severe reductions in military spending, removal of most (if not all) American troops from foreign bases, broadened powers for the EPA, enhanced Wall Street regulations, and a more progressive tax code. The Occupy insurgency was able to maintain a presence in numerous cities for weeks and even months, spreading even to Europe, while promising a more durable presence than has been the case with other movements. Given the deep popular anger over elite privilege, along with the steady rightward drift of the Republican Party, a citizen-based insurgency of this sort was to be expected. However, the question here (as before) is whether this movement could take social protest to the level of political strategy along lines of the Populists, who built a political party with specific programs, a list of candidates for government office, and a clear orientation to state power.

Environmental movements, first surfacing in the later 1960s, never generated a coherent political strategy like that of the West German Greens, as many ended up seduced by interest group reformism while others grew isolated from the political terrain altogether.[8] The deep ecology current produced a loose network of bioregional movements inspired by a fetishism of nature, emphasis on local resources and self-sufficiency, and embrace of community localism fitting the pursuit of ecological sustainability. Influenced by the communitarian legacies of Rousseau and Kropotkin, bioregionalism emphasizes reorganization of society through public sharing of land and its resources. Peter

Berg, founder of the San Francisco-based Diggers in 1966, later set up the Planet Drum Foundation as a clearinghouse for hundreds of bioregional groups spread across the United States. As we have seen, dozens of other grassroots environmental organizations have appeared in recent decades, many influenced by deep ecology and bioregionalism: citizens' groups dedicated to urban revitalization projects, tree planting, river and bay keepers, protest against toxic wastes, animal rights, water and general resource preservation. Thousands of urban farms currently thrive in every major American city, attuned to community life, self-sufficiency, local food production, and improved personal health. Such farms often meet fierce resistance from developers and city governments but have mostly still managed to preserve their domain. As of 2011, these and kindred local organizations remain firmly outside the political system.

Local organizations and projects have long dominated the environmental movement, based in a combination of interests: sustainability, social equity, community empowerment, urban revitalization, alternative energy sources, education for ecological consciousness. The Global Justice Ecology Project, based in Vermont, is dedicated to building bridges across many of these issues in a way that links grassroots struggles with environmental research and education. Another Vermont-based project, the Institute for Social Ecology, promotes similar community-oriented activities, merging local efforts to transform agriculture, technology, and urban life with critical pedagogy revolving around course offerings, research, workshops, and conferences. Other local organizations include Frances Moore Lappe's Small Planet Institute (in San Francisco), the Global Footprint Network (Oakland), Environmental Action (Boston), Center for Progressive Reform (New Mexico), the Oakland Institute (Oakland), and Center for Sustainable Organizations (San Francisco). Green America and Global Exchange have collaborated to sponsor a series of Green Festivals, nonprofit operations committed to harnessing the power of investors, businesses, consumers, and educators behind grassroots-driven environmental reform. The Festivals bring to the public dozens of exhibits for such projects as alternative technology, organic gardening, eco travel, green lifestyles, local arts and crafts, environmental careers, and responsible (green) investing. Partnerships are created with hundreds of local farms working for community empowerment, organic foods, and healthy lifestyles. Similar grassroots efforts proliferate around the world. Surely new waves of local mobilizations represent an essential building block for any future ecological revolution. For the most part, however, these projects—fascinating and appealing as they are—remain comfortably distanced from politics, especially in the United States, where the gulf between local activism and strategic leverage remains as wide as ever.

The prevailing tendency within American liberalism, noted earlier, has been the proliferation of interest groups seeking influence and change within the liberal state—the Sierra Club, World Wildlife Federation, Humane Society, Environmental Defense Fund, and dozens of others. Such lobbies focus on building membership bases, soliciting money, and shaping legislation with an eye toward policy reforms, made possible by a phalanx of well-paid organizers, spokespeople, and lobbyists. Oriented toward political intervention, such groups usually push single issues (or clusters of issues) to be addressed by Congress, state legislatures, government agencies, and the courts. None are interested in questioning much less overturning the existing power arrangements or pressing broader agendas such as military conversion or environmental sustainability—aside from ritual deference to those agendas. Some, as we have seen, gladly accept contributions from business interests attached to the corporate-growth apparatus.

A far smaller number of radical environmentalists embrace variants of direct action—understood as immediate *tactics* rather than political strategy—in response to pressing issues. Activists in Greenpeace have mobilized against nuclear testing and strip mining; in Earth First! against destruction of old-growth forests; in antinuclear groups against nuclear reactors and waste-disposal difficulties; in the Sea Shepherd Society against international whale hunting; and in the Animal Liberation Front against laboratory testing of animals at universities and elsewhere. One well-known form of direct action has been "monkey wrenching"—putting human bodies in the path of corporate developers—taken up in earnest by Earth First! in northern California. Small but influential, such groups have an affinity for counterculture values and spontaneous actions while retaining movement distance from establishment politics. Direct action protests are a defining *modus operandi* of many animal rights groups, their anger directed toward the widespread (and usually unnecessary) use of animals in scientific and medical work. Embracing a strong moral righteousness, such groups are normally condemned to pre-political marginality and impotence.

One popular ecological approach best described as "new age" is grounded in a combination of spiritualism, faith in positive human outcomes, devotion to cosmic harmony, and rejection of "politics" as a domain of corruption and futility—an outlook most compatible with deep ecology and such tendencies as eco-feminism. Change stems from personal experience and consciousness transformation, integral to a painless "revolution" like the one described long ago by Charles Reich in *The Greening of America*.[9] More recently, Paul Gilding, in *The Great Disruption*, argues that ecological revolution is already a mature process fueled by cultural and spiritual enlightenment that has already been sweeping American society, as human beings move organically toward a "higher evolutionary state," writing, "It is the crisis itself that will push humanity to its next stage of development and allow us to realize our evolutionary potential. It will be a rough ride, but in the end we will arrive at a better place."[10] According to Gilding, humanity is now at the point of moving to the "next stage" thanks to the positive energy and good will of people, governments, and corporations desperately wanting to avoid global crisis and collapse. Slash global energy consumption and fossil-fuel reliance? Millions of world citizens are already involved in this process, using new leverage through developing market and technological possibilities. A former Greenpeace activist who moved into higher corporate echelons, Gilding believes a dramatic shift in values consistent with an ecological outlook is already taking place—mainly unfolding *outside* the dominant public sphere. Building on the "extraordinary capacity of human ingenuity," people are innately driven to do the right thing: "With all of us in charge, we vote every minute of every day" (presumably as consumers).[11] Here again we have a perspective on change that in principle steadfastly avoids the mundane realm of politics.

On the premise that "we are the system," Gilding believes that corporations are prepared to respond to the will of an awakened people—that is, to a "gradual change in values" as attitudes of peace and cooperation come to replace the aggressive pursuit of economic self-interest. Typical of such new-age fixation on altered consciousness, what drops from view here is historical or structural analysis that might inform effective political action; change unfolds organically, the result of individual psychological conversions on a large scale. Like Gilding, Deepak Chopra foresees a gradual, presumably inexorable enlightenment process in which collective humanity is moving toward ecological awareness, leaving behind destructive traits of narcissism, consumerism, and violent aggression. For Chopra, this process is eminently spiritual and personal although, consistent with new-age thinking, he remains vague about mechanisms and agencies of imputed human progress. Empirical indicators of such transformation are nowhere to be found. Looking at the American experience, Chopra finds a nation perpetually at war with society, the world, and nature—a population of atomized, lost beings desperately seeking false solutions through consumption, technology, and other elixirs. The result is a system that perpetuates war, terrorism, poverty, and destruction of all sorts. In this setting, however, Chopra finds grounds for optimism and progress, but how and under what conditions the new spiritual awareness is supposed to emerge (and flourish) remains a mystery.

In similar vein, Frances Moore Lappe, writing in *Eco Mind*, argues that the ultimate basis of reversing the ecological crisis is an emergent human alignment with nature best characterized as a collective "eco mind."[12] Though Lappe, to her credit, situates this transformation in the concrete growth of local environmental and other progressive groups, for her the centerpiece remains diffusion of a new popular worldview on questions of wealth, growth, and prosperity. In the end, the power of human consciousness is decisive: "So why not put the power of ideas to work enabling life on earth?"[13] Following Gilding and Chopra, Lappe exudes unwavering optimism in the role of popular beliefs feeding into an "immensely liberating ecology of hope."[14] The Global Coherence Project embraces this same ecological vision, foreseeing a new era of peace and cooperation in a "new-age world" made possible by a dramatic shift in world consciousness that is already well advanced. One of the project's leaders, Gregg Braden, anticipates a future marked by spirituality-based change linked to expanding positive energies in the "earth's field." Like the others, Braden sets forth a vision strong on hope and optimism but largely devoid of politics.

Optimism in evolutionary, peaceful yet radically transformative change has spread with the popularity of new technology, viewed by many community organizations, social movements, and interest groups as the wonderful gift of human progress, something of a panacea for solving urgent problems. Young, well-educated activists embrace online communication, blogging, social networking, and such devices as smart phones as the political wave of the future, a source of virtual interaction, electronic democracy, intellectual dissent, and information access denied within mainstream politics and media. Used diligently and creatively, technological venues can stimulate participation, foster organizing, bring people into the streets, and perhaps even topple dictatorships—as they have done in Egypt, Libya, and other locales of the 2011 "Arab spring" energized by a dramatic wave of mass upheavals. The "twitter revolution" takes anti-system opposition to new levels of human potential. In the United States, organizations like MoveOn—with some three million members, by 2010 the most visible progressive organization in the country—celebrate the virtues of online activism, with its ongoing electronic town meetings, cyber polls, civic networking, webinars, blogging, and other modes of technological citizenship. For these groups technology emerges as far more than a tool of political communication: it constitutes the source of an entirely new model of community and democracy, an indispensable form of modern social engagement. The apparent breakthroughs of both the Arab revolutions and Wall Street Occupied protests have endowed such technological utopianism with new credibility and appeal.

Leaving aside the degree to which the Arab "revolutions" actually overturned the old power structure (beyond ridding the societies of individual tyrants), the American experience offers abundant cause for skepticism regarding such claims. Like all modes of communication, social networking, twitter, and other online devices will be important to any human activity, including movements for change. One immediate difficulty with the more utopian views is that the ruling interests have full access to precisely the same technology and indeed possess enough resources to make even *greater* use of it for sophisticated methods of control through government, corporations, law enforcement, the military, the media, and intelligence agencies. It might be argued that, in general, the new technology does more to bolster than to subvert networks of domination in an era when individual privacy and autonomy have been systematically undermined.

The deeper question, however, pertains to whether online activism can encourage and sustain a vibrant radical *politics*—or whether the technology is likely to reinforce depoliticizing tendencies already well advanced in the industrialized world. Do organizations like MoveOn in fact represent the cutting edge of transformative change in a fashion that opens new space for popular insurgency, including efforts to reverse the global crisis? Perhaps—but the evidence to date is far from convincing. Insofar as politics entails a struggle for power connected not only to grassroots involvement but to political strategy and party formation, the new technology falls considerably short, in part because of its own logic. Evidence to date suggests that online technology, while facilitating personal communications, tends to diminish civic interaction and citizen participation as real-life forms of *public* activity. Face-to-face engagement vital to durable community organization can easily give way to more anonymous, impersonal, episodic forms of interaction. There is a certain self-absorbed, fragmented dimension to electronic participation, not entirely consonant with deep citizenship or social change. Within online culture, moreover, there is actually widespread disdain for a politics that does not fit conventional institutional and ideological norms. Groups like

MoveOn, whatever their achievements, rarely address questions of how political strategy might channel popular anger, frustration, and strivings into coherent alternatives to the status quo. Wedded to an instrumental, tactical outlook, therefore, it is hardly surprising to find that MoveOn and kindred organizations have expressed little interest in moving beyond the Democratic Party orbit. New technology provides rather comfortable avenues of public access, information gathering, and personal interaction, but in the United States at least its great transformative potential remains a distant vision. It has not yet embellished a much-needed recovery of politics—and might well do more to counter it.

Other barriers to a coherent movement fusing ecological priorities with those of labor and communities persist, especially in the American setting. Rare have been the occasions when the labor movement and environmentalists have converged around a specific political challenge; what provides jobs and benefits to workers might well run counter to environmental agendas. One case in point revolves around a rock-quarry project near Temecula, California—a massive project deeply embraced by labor but fiercely opposed by community and environmental groups. The quarry, a 414-acre landscape to be exploited by Granite Construction, would provide hundreds of jobs over several decades but would destroy pristine rural areas and deliver heavy toxic pollution to the Temecula region. The local band of Luiseno Indians argued that the Quarry would degrade very sacred land. A city-sponsored health study reported that hundreds could die from toxic pollution, with the health of perhaps thousands more severely impaired. As of early 2012 the struggle remained at an impasse.[15]

As the ecological crisis worsens, therefore, a recovery of politics—the struggle for deep citizenship, social change, and radical governance—is not yet on the horizon. This has all the makings of historical tragedy, as such recovery takes on added urgency at a time when the gulf between community groups and dispersed social movements, on the one hand, and effective political leverage on the other seems wider than ever.

The Global Imperative

One article of theoretical faith among early Marxists was that socialist transformation, to be sustainable over time, would have to unfold as an *international* project, cutting across national borders that help perpetuate class power. For Marx and Engels, of course, the proletariat was expected to develop class rather than national consciousness, rejecting any political (or military) allegiance to the domestic bourgeoisie and its state power. At the time of the Bolshevik revolution Leon Trotsky's concept of "permanent revolution" approached the transition from feudalism to capitalism as (ideally) building toward a socialist revolution driven by widening insurgencies in Europe and beyond. The very idea of "socialism in one country," later associated with Stalin, was implausible—indeed a contradiction in terms. Socialist politics would inevitably be undercut by a resurgent nationalism, which turned out to be the case for Soviet development. Conquests of state power in single countries at a time when international capital retained its hegemony would, sooner or later, lead to isolation and then collapse. Today, at a time when the power of transnational capital dwarfs that of a century earlier, the global imperative stressed by Trotsky seems more (not less) compelling. Further, like capital itself, the ecological crisis respects no national or territorial boundaries, being truly planetary in scope. The nation-state as such constitutes a potential obstacle to global solutions as it is drawn to national economic competition, resource conflicts, geopolitical struggles, and (notably for the United States), outright warfare. The most ambitious domestic attempts at change, therefore, will ultimately demand international collaboration, for such attempts can easily be neutralized by capital flights and other subversive challenges from abroad. This applies dramatically to the problem of climate change, all the more so in the wake of unproductive environmental summits at Rio, Copenhagen, and Cancun.

Beyond these issues, international partnerships hold obvious advantage in terms of political efficacy—a maxim clearly understood by transnational capital, which possesses an array of its own global integrative mechanisms, including the World Bank, IMF, WTO, and European Union. While local and national strategies are of course indispensable—indeed a necessary point of departure—they remain vulnerable to the pervasive reach of both national and world capital. Recognizing this challenge, the first socialist parties created the Second International in 1891, based mainly in Europe but dedicated to the goal of international working-class cooperation. This was largely a fiction, however, since social-democratic leaders held to a strong nationalism that was embarrassingly revealed at the outset of World War I when the parties, with few exceptions, aligned with their own national military forces in the armed conflict. Eventually superseded by the Socialist International (SI) in 1952, which expanded to 54 member parties with some 15 million members by the 1970s, the network bequeathed a legacy of tepid reformism with national parties only loosely bound together, their connections weakened by decades of class collaboration and political institutionalization. Today the SI parties are essentially historical agencies of secular liberalism, allied firmly with capital as well as their specific nation-states. As for the Communist (Third) International spawned by the Bolshevik revolution in 1919, its member parties (typically weak and isolated) were fully subordinate to the Soviet Union until the organization was disbanded in 1943, part of Stalin's effort to placate World War II allies. The success of a Leninist revolution in Russia conferred on Moscow "leadership" of world Communism, codified by the famous 21 points binding national members and enforced by the Soviet party-state during the 1920s and 1930s. Noteworthy here is the fact that the epic twentieth-century Communist revolutions—in China, Vietnam, Yugoslavia, and Cuba—succeeded largely independent of Soviet power, as they were overwhelmingly fueled by nationalist mobilization.

Sequels to the largely failed Second, Third, and Socialist Internationals have been numerically few and politically marginal. A "Fourth International" founded by a small nucleus of European Trotskyists never transcended its own sectarian isolation. The European Green Party, as mentioned, was founded in 2004 as a loose federation of more than 40 national parties—theoretically the start of a much larger Green International. In 2001 activists from around the globe, inspired by the Seattle protests, organized the World Social Forum in Porto Alegre, Brazil, which turned into an annual gathering of thousands of participants committed to a democratic, egalitarian alternative to global neoliberalism. The Forum established a Charter of Principles oriented toward non-corporate globalization that delegates from dozens of countries were able to endorse. The second WSF meeting attracted some 12,000 delegates from 123 nations, with 80,000 attendees and nearly 700 workshops. By the fifth gathering in 2005, held again in Brazil, the number of participants swelled to 155,000. Regional forums have been organized in Europe, Asia, Africa, Latin American, and the United States, all dedicated to some variant of radical change—although popular movements have become less visible within the WSF orbit than the more conservative NGOs. While serving as a vital axis of global anti-system tendencies, the forums (as the label implies) have no defining political mission like that of the earlier internationals; there is no WSF party or even strategy, nor is that much discussed. Committed to "reinventing democracy," the forums have evolved as mostly intellectual and cultural assemblages—a laudable but nonetheless pre-political enterprise.

As for the Greens, the European grouping of parties does aspire to a worldwide presence, though somewhat loosely integrated. While the Greens have maintained a strong international presence, their greatest influence is centered in the industrialized North, still inspired by the original German model. In contrast to the WSF, the European Green Party advances a clear political ideology and strategy: sustainable development, limits to growth, social conversion, demilitarization, nonviolence, a merger of electoral activity and grassroots struggles directed toward institutional power. While marginalized in most countries, sometimes nearly invisible, Green parties on the whole made impressive strides in the years following the West German breakthrough, reflected in the birth of the European Party, which in 2011 returned 46 deputies to the European Parliament. Aside from the aforementioned ideological divisions, crucial questions arise, mostly relating to the capacity of national parties to win enough popular support for governmental leverage. There remains the great challenge of North-South relations: from the outset Green politics has been a difficult sell to less-developed nations preoccupied with their own economic growth. As of 2011 no expanded Green International was yet in sight. Another central question is whether the Greens, in or out of government, can resist the pull of deradicalization as they go about building domestic and global support. Finally, there is the matter of how any future Green (or other) worldwide political collaboration might solve the old dilemma of preserving unity in diversity—what Italian Communist leader Palmiro Togliatti called "polycentrism" in the early 1960s. Answers to such questions could go far in determining the human capacity to reverse the headlong advance toward ecological catastrophe.

An Ecological Growth Agenda

The global crisis raises a new crucial questions regarding the familiar capitalist regimen of "economic growth"—a discourse central to neoliberal ideology as well as to environmental movements fixated on "limits to growth" and even "no growth." Both modalities are riddled with fictions, myths, and distortions. An emergent ecological worldview punctures mystifications about "growth" as it affirms the vision of a qualitatively new (non-wasteful, nondestructive, sustainable) model of human progress. Presently the motif of "growth" endlessly trumpeted by corporate elites and their economists represents nothing so much as a cover for uneven development leading to further exploitation, social inequality, and environmental ruin. For the ruling interests, in fact, "growth" amounts to a substitute for broad progressive initiatives, serving more as a legitimating ethos for policies that favor oligarchical power and a tiny super-wealthy stratum, offered as something of a panacea like promises of fiscal stability, "peace," and "family values." Conversely, calls for drastically reduced growth or "no growth" by environmental theorists and activists—championed by deep ecologists—not only impede popular receptivity to ecological politics but perpetuate the very myths surrounding neoliberal agendas. A new standpoint is long overdue, advancing a schema of ecological development tied to social equity, peaceful relations among nations, democratization, and sustainability, in turn opening the door to qualitatively new ways of measuring economic outcomes and societal wealth. We need infinitely better ways to assess exactly what is produced and consumed—as well as *how* it is produced and consumed.

In fact is that present "growth" regimen contains vast levels of waste, destruction, inequality, and imbalance, only thinly veiled through misleading references to "free markets," "free trade," and aggregate measures of Gross Domestic Product (GDP). This is a system where "growth" has little relationship to the general quality of life, measured by levels of education and housing, health and medical care, democratic access, work relations, infrastructure quality, and natural resource supports. Standard economic indicators focusing on simple quantitative "growth" and material abundance do not measure societal living standards—that is, real indices of human progress—since they are derived from outmoded means of calculating the production and distribution of goods. By stripping away the (false) relationship between corporate-defined "growth" and social well-being, an ecological outlook points toward heightened living standards while also dramatically reducing GDP and, with it, negative human footprints on the global habitat. The reality is that existing measures of "growth" conceal untold amounts of waste and destruction in the resources consumed by corporate superprofits, a lopsided emphasis on "private" over "collective" forms of consumption, a grossly inefficient energy system, a militarized economy, a meat-based agriculture and fast-food system, and a top-heavy finance capitalism. Viewed thusly, "growth" under modern transnational corporate auspices is fraudulent, nowhere more so than in the United States, where the neoliberal growth ideology has been raised to the level of a civic religion; while generating increased profits and wealth for a small elite, "growth" is now tantamount to elevated levels of social, political, and environmental impoverishment on a world scale.

How and where does the system allocate resources in accordance with the corporate-engineered "growth" mandate? The first point is that, with wealth concentrated in such few hands, resources are not equally available throughout the general population: GDP does not even pretend to measure precisely how economic goods and services are distributed across the entire society. The enormously skewed distribution of wealth, income, and investments in the United States is perhaps the major source of waste and distortion within a corporate-growth system that does not provide adequately for tens of millions of its citizens. In 2010 the top 1 percent of Americans owned no less than 35.6 percent of total wealth, 43 percent of financial assets, and 49.7 percent of total investments while the bottom 80 percent had just 7 percent of wealth and one-fourth of all households possessed zero net worth. The leading 500 income earners *averaged* $350 million each and enjoyed net worths of at least one billion dollars apiece, a collective treasure trove of more than $1.5 trillion—8 percent over 2009 levels, even in the midst of severe economic downturn and growing misery for the vast majority. The number of American households with resources over one million dollars grew to nearly eight million in 2009, those over five million dollars to roughly one million. The ratio of CEO/top executive pay to that of average workers skyrocketed to an obscene 400:1 ratio in 2010. These amassed fortunes at the top of the class pyramid, when juxtaposed to growing scarcities for the majority—not to mention unemployment, home foreclosures, bankruptcies, and shrinking public services—reveal more clearly than anything the outlandish myths and distortions endemic to the "growth" agenda. The obscene and mostly parasitic wealth of a relatively few elites, when spread out more equitably over ten years, could eradicate poverty, pay off all outstanding home mortgages, provide health coverage for millions of uninsured for several years, finance green public transit networks in at least 20 major cities, and buy housing for every homeless person in the country. Spanning a decade or more, such an equitable shift in wealth distribution could help transform American society along far more economically, socially, and ecologically sustainable lines.

Further, the corporate-growth economy measures within its total GDP huge sectors of privatized, largely nonproductive wealth, using the same quantitative values applied to more socially useful, public forms of consumption. Official economic statistics do not reflect profound differences in what gets produced and consumed, much less how it gets distributed. In 2011 the United States had a world-leading GDP of $14.7 trillion, with domestic financial assets of nearly $131 trillion. What percentage of GDP is wasted in manifestly nonproductive activities such as insurance, advertising, and lobbies—or in a hyper-commercialized agricultural system, the permanent war economy, a gluttonous fossil-fuel energy system tied to the auto culture, privatized medical-pharmaceutical complex, a financial plutocracy generating little actual value despite accounting for more than 10 percent of total GDP? There is of course no simple calculus for such routinized economic waste, but it surely accounts for several trillion dollars yearly—perhaps even half of existing GDP.

A protracted shift toward more equitable, productive, and sustainable modes of economic activity could easily reduce "growth" by one-third or more yearly at even higher standards of living for the general population (while imposing little if any hardships on the wealthy few). Thus a departure from privatized medicine (accounting for 16 percent of GDP, or $2.6 trillion) in favor of socialized model would save more than one trillion dollars in profits, excessive costs, overlapping bureaucracies, unnecessary treatments, and massive drug expenditures alone. A similar move toward green (non-carbon) public transportation systems, allowing for an initial one-half reduction in auto use, would save additional trillions of dollars in lessened reliance on fossil fuels, greater transit efficiency, diminished air pollution, and better allocation of space. A scaled-down military, quite feasible in a world largely bereft of national enemies, could save more than a half-trillion dollars yearly, freeing those resources for public transit, housing, education, and infrastructure revitalization. Imposing a strict cap on the pay and bonuses received of CEOs and other high executives would save yet more hundreds of billions.

The U.S. military, of course, is a uniquely wasteful and destructive sector of the economy, with a 2011 budget of more than a trillion dollars allocated to hundreds of bases scattered around the world, a massive intelligence apparatus, high-tech weapons programs, nuclear development, R&D, veterans benefits, and so forth—not counting more hundreds of billions yearly for wars abroad. Pentagon spending is expected to rise to as much as $1.5 trillion by 2014. At present Washington accounts for nearly *half* of all global military expenditures, roughly six times that of China and 12 times that of Britain, France, and Russia. The military budget amounts to 28 percent of federal tax revenues and 51 percent of all national discretionary spending at a time when budget deficits are depicted as a threat to national survival. Pentagon spending accounts for about 5 percent of GDP, far beyond that allocated by any other country. Here one is prompted to ask, what possible harm to U.S. national security could result from, say, armed forces cutbacks to a level of $300 or $400 billion? What might be done with the savings, for even one year? Such questions naturally have no resonance within conventional GDP discourse—or the corporate media.

To say the banking industry constitutes a bottomless reservoir of economic waste is a gross understatement: at 10 percent of GDP, finance capital dominates the economic landscape as a predatory monster, deeply involved in every sector. In 2010, banking institutions reaped fully 30 percent of all domestic profits for a sector that contributes no genuinely productive goods and services while burdening other sectors with risky investment schemes, high interest rates, commodity speculation, and the exporting of industries and jobs to China and elsewhere. Accounting for a flow (in 2011) of nearly two trillion dollars globally and domestically, financial profiteering continues to distort outcomes and threaten the American economy, with few strict regulations, vastly undermining the official value of GDP. At present banks are extracting massive rents, interest, and profits from a system in which productive industries are captured by financial schemes that, as we have witnessed, lead to misery for the general population. Meanwhile, the banking giants continue to expand and continue to fight public regulations. Although banking services are obviously indispensable to any economy, the introduction of a non profit public financial system to replace of the Wall Street juggernaut would save easily a trillion dollars in waste yearly.

Energy consumption represents yet another zone of great economic waste and destruction. At present the United States spends many trillions of dollars annually on all forms of energy, while still mostly reliant on extremely costly and harmful fossil fuels for virtually every realm of human activity. Leaving aside the relatively small nuclear sector, carbon-based energy is by far the most costly source and still fuels transportation, communications, agriculture, public infrastructure, and the military. The United States (along with China) currently sits atop the world energy-use pyramid, consuming five times the level of Japan, seven times the level of Germany and France, and 12 times the level of Italy—all industrialized nations. With the United States in the lead, the world has consumed more total energy over the past *two decades* than previously throughout the entire industrial period. Since 1980 the United States has averaged 350 BTUs per capita, compared with the global level of 72 BTUs, about *five times* what is consumed (on average) across the world. As of 2011 the total U.S. energy consumption was as follows: 40 percent oil, 23 percent coal, 23 percent natural gas, 8.4 percent nuclear. Alternative sources (solar, wind, etc.) accounted for a mere 7.3 percent, a level predicted to increase only slowly in coming years. The United States consumes roughly 20 billion barrels of oil yearly, compared with 4.4 billion for Japan, 2.5 billion for Germany, and 1.9 billion for France. Indeed the American military alone, with its fleet of 200,000 vehicles and hundreds of bases around the globe, uses fully 1 percent of total U.S. energy resources—more than all but 35 countries in the world. What percentage of U.S. resource utilization—including not only fossil fuels but land and water—could be saved through more ambitious embrace of alternative energy? Change, of course, would eventually require much deeper institutional and ideological transformations. Savings here as elsewhere cannot be easily calculated as it depends on so multiple variables, not least being changes in transportation, agriculture, and the military. Lappe, for example, suggests that as much as 87 percent of existing U.S. energy consumption is squandered.[16] Even if we conclude that 60 to 70 percent is wasted, the savings in wealth and resources *yearly* would surely reach into the trillions.

The world food crisis raises pressing questions about the glaring dysfunctions and inefficiencies of modern agricultural production, a system dependent on exhaustive use of natural resources (land, water, fossil fuels) for large-scale agribusiness geared overwhelmingly to meat and dairy products. As of 2011, there was little indication of a shift in global food production toward sustainability, although proliferation of local urban farms is cause for some optimism. Unfortunately, world meat and dairy intake continues to grow rapidly in most regions, keeping pace with industrialization and gravitating toward "modern" diets, including fast foods. Feeding grains to more and more animals has created an increasing drain not only on food stores but on available water, land, soil, and fossil fuels, with no end in sight. The stupendous waste endemic to growing, fertilizing, processing, and transporting foodstuffs is manifest across the globe. Presently *one half* the earth's land mass is grazed by livestock, while the amount of U.S. cropland devoted to livestock feed is more than 60 percent. The total American grain production consumed by livestock is 70 percent. It takes roughly 16 pounds of grain to generate one pound of food from beef or chicken, essentially wasting 90 percent of foodstuffs in a world hurtling toward excruciating food shortages. Fossil-fuel calories needed to produce simply one calorie of protein from meat amount to 80, compared with just two for grains or soybeans. In many areas of the world (including most of the United States) livestock production alone drains half of all water consumed, as beef requires more than 5,000 gallons of water per pound compared with about 25 gallons for tomatoes, lettuce, potatoes, and wheat. Even a modest global shift toward vegetarian diets (say 10 to 20 percent) would generate vast savings in diminishing resources—not to mention improved health (further savings) and dramatically reduced greenhouse emissions (yet more savings).

Even modest reductions in the deep systemic waste and destruction of the American economy—spanning finances, health care, transportation, energy, agriculture, the military—could permit far-reaching cuts in GDP while supporting markedly *increased* living standards for the general population. More protracted and deeper changes would naturally generate more durable change toward economic rationality, social equity, and environmental sustainability. From this standpoint, GDP turns out to be a worthless indicator of wealth, well-being, and even growth. In settings where resources are more unevenly distributed present GDP levels are even more misleading. The corporate-state obsession with "growth" is therefore nothing but a delusion, meaningful only as an elite scheme for superprofits and legitimation. Further, the oft-repeated choice between "growth" and "no-growth" is a mirage insofar as it fails to critically assess the actual character of GDP that is now based on false quantitative, aggregate measures. Human needs can be abundantly met with much fewer resources (perhaps half of existing levels) while significantly reducing the destructive societal footprint on global ecosystems. In the end, the great utopia of abundance promised by both the Enlightenment and capitalism winds up a fraudulent trick, turning into a source of public squalor, plutocracy, economic waste, militarism, and environmental disaster.

The Logic of Deradicalization

A central motif of this book is that the historical struggle for a socially and ecologically transformed world—for a livable, just, sustainable planet—must achieve radical political articulation at a time when the modern crisis threatens to veer out of control. Radical change by definition involves the spread of anti-system discourses and practices, going beyond the limits of earlier liberal, anarchist, Jacobin, and social-democratic models of change—dependent on a shifting balance of forces in society as a whole. A new strategic model flows from a synthesis of party and movements linking national, local, and (eventually) global arenas of change that, as we have seen, calls forth the initial Green schema in West Germany. Mounting attention to the electoral side, favored by Realos, posed the threat of deradicalization like what had previously overtaken the Socialists, Communists, and kindred parties dedicated to the parliamentary road; Green strategy, it turned out, offered little insurance against such corrosive pressures.

An original architect of the parliamentary road, Bernstein theorized a socialist *modus operandi* that seemed eminently practical, an evolutionary, peaceful transition seemingly congruent with the flow of history. Early social-democratic politics assumed that (a) the working class would become a majority force with advancing industrialization, (b) trade union reforms would bolster the economic (and thus political) influence of labor, and (c) electoral gains would transform workers into a decisive (majority) political force, allowing for the socialization and democratization of liberal-capitalist institutions. For Bernstein, a workable socialist strategy meant breaking with the main tenets of orthodox Marxism: a science of history, the crisis scenario, immiserization of the proletariat, capitalist breakdown leading to insurrection. Socialism was expected to grow organically within the very matrix of liberalism-capitalism, building on citizen participation within the nation-state as sectors of the middle strata joined workers in a broad multi-class alliance. This model naturally appealed to diverse social groups—workers immersed in daily struggles, trade union leaders involved in contractual bargaining, party leaders seduced by bureaucratic privilege, elements of the middle strata looking for a more "humane" capitalism, intellectuals grateful for a nonviolent route to change. Unfortunately, no social-democratic party following the Bernsteinian approach ever managed to escape the incessant logic of deradicalization, what Robert Michels called "embourgeoisement."[17] The parties served as vehicles of a capitalist welfare-state driven by expansive social priorities, Keynesian fiscal and monetary policies, and institutional regulation of class conflict. Bernstein's fanciful vision of a gradually maturing socialism spreading across Europe and the world rested on an illusion that anti-system change could take hold naturally out of the very processes of capitalist development.

History resoundingly negated Bernstein's model of socialist transformation, as the deradicalizing thrust of a primarily electoral strategy was visible already during Bernstein's lifetime and the supposed heyday of European social democracy. Such failure would be extensively analyzed in the work of such writers as Michels, Carl Schorske, Peter Gay, and Guenther Roth.[18] These critics argued that party and trade union leaders became closely allied with their national bourgeoisie, pushing social democracy along the comfortable route of moderation and institutionalization; socialist identity might be retained in theory but was jettisoned from the sphere of everyday politics. This pattern took hold in the early years of the Second International and would become consolidated across subsequent decades, as social-democratic parties and governments morphed into variants of Keynesian state-capitalism in Europe and beyond. Moreover, later Socialist and Communist parties pursuing the same electoral strategy—often called "structural reformism"—suffered the same fate, including those that identified with a "third road" between Western capitalism and Soviet-style Communism. Mass-based Communist parties in Italy, France, and Spain, often winning strong bastions of local power, experienced a similar deradicalizing outcome during the post–World War II era, extending to the grand promise of Eurocommunism in the 1970s.[19] Socialist parties swept into national power during the early 1980s—dramatically so in France, Spain, Italy, and Greece—but quickly succumbed to the pressures of capital, adopting the very austerity policies championed by Margaret Thatcher in Britain and Ronald Reagan in the United States. In each case party leaders followed a parliamentary strategy in which popular movements were compromised for the sake of electoral gains. Embracing norms of liberal-democracy (or looking to *broaden* them), Socialist and Communist parties became fixated on legislative accommodation, interest group bargaining, bureaucratic leveraging, and government patronage—scarcely a formula for radical change. Where Socialist and Communist parties won power—usually in coalitions—they always departed from their original identity, choosing partnership with the ruling interests as something of a *fait accompli*. One dilemma for oppositional politics, today as in the past, thus revolves around how to sustain radical goals through the inevitable compromises and setbacks of electoral activity—a dilemma sharpened by realization that no truly *democratic* process of social transformation can occur where the strategy is to either smash (Leninist) or totally abstain from (anarchist) the liberal state.

The historical dynamics of radical opposition—whether anarchist, syndicalist, or council communist—is that popular struggles must be centered *outside* liberal-democratic institutions, based in some combination of local movements, workers' councils, community assemblies, and direct action. The legacy of "dual power" aligned with mass spontaneity and local self-management has generally been at the core of this strategy which, left to its own momentum, leads to political impasse. Such anti-system impulses have in fact been revived within advanced capitalism, as in the case of the new left and (in Europe) extraparliamentary opposition during the 1960s, followed in the 1970s by the spread of new social movements and rebirth of anarchism. What these nonelectoral insurgencies held in common was dedication to grassroots self-activity, hostility to conventional politics, and deep skepticism of the traditional left, whether social-democratic or Leninist (often dismissed as "Stalinist"). They were advocates of direct action, civil disobedience, social upheaval, and countercultural lifestyles. Framed against the backdrop of the modern ecological crisis, this legacy—with all its limits and flaws—most closely fits the contours of a radical green strategy.

Electoral politics has generally, but not always, clashed with grassroots movements—the one firmly embedded in dominant structures and routines, the other in opposition to or overthrow of those very structures and routines. An ecological revolution is unthinkable without the latter: popular struggles against the corporate state can resist the deradicalizing logic of elections, parties, legislative bargaining, and bureaucracy. If structural reformism means business (and politics) as usual, radical strategy challenges such normalcy, building upon the unruly outpourings and disruptions of mass insurgency. Confined to normal structures, laws, practices, and discourses, the electoral path when pursued alone guarantees political integration. An ecological politics with hopes of reversing the crisis must ultimately break with an institutional order embedded in the ravages of destructive, unsustainable neoliberal development. At the other extreme, rebellious forces outside the electoral arena historically *avoided* challenging state power, with predictably depoliticizing outcomes. Today, what might be labeled "social-movement anarchism" (Mexican Zapatistas, "black block" global justice activists, green street insurgents, etc.) have neither expanded their civil society presence nor carried forward an efficacious strategy for radical change. Such groups insist, with some validity, that direct participation in liberal state serves to legitimate the corporate-state system and all it represents—but their abstention has been purchased at the steep price of political marginalization.

If one-dimensional strategy harms radical democratic prospects, the original Green model would seem to offer a way out, based as it is on a convergence of party and movements, electoral and grassroots opposition that, in theory, might resist the twin threats of deradicalization and isolation. To date, however, no Green organization has managed to fight off these threats which, in the case of larger parties, has meant adaptation to the pressures of social-democratic reformism. While no simple blueprint exists (or could exist) for preserving a dual trajectory, new social conditions could well give rise to more hopeful outcomes. Mounting ecological challenges, combined with material and social crises, could force change-oriented movements and parties leftward, opening wider space for radical alternatives. Normal politics will be more difficult to ritually sustain at a time when "normalcy" is associated with chaos and breakdown. Forms of "dual power" have often flourished at moments of aggravated crisis: Russia before and during World War I, Italy just after the war, Spain in the late 1930s, France in May-June 1968. A crucial question is whether such inherently unstable oppositional forms can persist over time—that is, whether they can be strategically durable. The answer so far has not been particularly encouraging. Deepening crisis, in any event, will surely improve anti-system prospects, allowing for broadened insurgency as more people choose to fight for a healthy, thriving, and balanced natural habitat along with better living standards. Yet crisis alone, as history shows, hardly ensures a leftward shift, especially in societies dominated by a powerful corporate media; as in the past, fascism or some equivalent could benefit from chaos and breakdown. Outcomes will naturally depend on the balance of social and ideological forces, on the extent to which ruling elites can preserve their hegemony in the face of new challenges, and on the capacity of insurgent forces to create movements and parties that can compete for governmental power.

A Nonviolent Radicalism?

The role of violence in social change has long posed dilemmas for political theorists, leaders, and activists—dilemmas, however, that nowadays rarely provoke debate given high levels of consensus on the virtues of nonviolence. Influenced by the legacies of Mahatma Gandhi and Martin Luther King, among others, leftist politics in the industrialized societies is shaped by an ethos of nonviolent resistance often linked to pacifism and tactics of civil disobedience. Debates over violence did of course surface historically, often involving Marxists and anarchists. It was Bernstein who promised a relatively harmonious, gradual, and *peaceful* road to socialism, with elections, union struggles, and cooperatives replacing the insurrectionary and implicitly violent "overthrow" scenario championed by Lenin and the Bolsheviks (and earlier by Marx). A philosophy of nonviolence thoroughly infused social movements of the 1960s and beyond—Civil Rights, the new left, antiwar, feminist, ecology, the counterculture. By the 1980s, as we have seen, the Greens advocated nonviolent politics not only as tactics but as philosophy, a way of life. More broadly, pacifism in the modern setting has emerged as a symbol of noble intentions, the sign of a morally righteous outlook that is rarely questioned by liberals and even progressives. After all, it had the unshakeable imprimatur of two great political figures of the twentieth century,

Gandhi and King. The decline of the Marxist left, most of which held out the necessity of violent revolution, only strengthened this ideological tendency.

If earlier leftist traditions like nineteenth-century Russian anarchism, syndicalism in the tradition of Georges Sorel, Bolshevism, and Third World movements influenced by Frantz Fanon and Che Guevara had often viewed political violence as not only imperative but cathartic, nowadays most everyone, regardless of ideology, pays homage to the notion that politics ought to be nonviolent even where conflict is sharp. To be sure, the idea that violence should be minimized during struggles for change has rarely been questioned—any more than hoping for an end to militarism and warfare. The crucial question here turns on the feasibility of a nonviolent *revolution* directed against firmly entrenched power structures that in themselves show repeated (often embellished) willingness to use deadly forms of violence to maintain their power and wealth. The conventional wisdom is that violence, whatever its context or motivation, simply begets more violence, thus rendering it counterproductive however politically "successful" it might be. Nonviolent resistance—noble, uplifting, defensible—is needed to prefigure a peaceful, harmonious system, and who would want to question time-honored lessons from icons like Gandhi and King?

While this outlook now appears universally shared as something of a virtue unto itself, it has similarly managed to avoid scrutiny or debate. Deeper philosophical and political questions are met with calculated silence, on the (unexplored) assumption that variables such as social *context* and political *objectives* are entirely irrelevant. At this point it is simply worth asking whether the global ecological crisis can be effectively confronted, much less reversed, by means of purely nonviolent struggles historically suited to reformist liberalism and social democracy. Do matters of urgency (time), global dimension (scope), and institutional power to be overturned ultimately alter the ideological terrain on which such matters of political ethics are understood? Does the everyday violence and destructiveness of concentrated power enter into the equation? Is radical insurgency even thinkable where politics is *ipso facto* limited to the parameters of nonviolent resistance? Further, what actual historical models of nonviolent social transformation exist for present-day emulation? The answer must be that, contrary to the received (and tightly enforced) political consensus, no such models are to be found: celebrated examples (Bernstein, Gandhi, King) never demonstrate what is often self-righteously claimed for them. A nonviolent anti-system strategy cannot be shown as adequate to the deep and urgent political challenges ahead. A closer look at *revolutionary* traditions reveals an uncomfortable truth for the apostles of peaceful change—that every historical instance of successful radical transformation has required, to varying degrees, uses of political violence. Surely the great modern revolutions (French, American, Russian, Chinese, etc.) were anything but peaceful celebrations. The birth of modern nation-states (the United States, Japan, Israel, South Africa, Algeria, Vietnam, Cuba, etc.) has typically been marked by bloody upheavals, foreign intervention, and civil wars. Are there any logical reasons to suppose that a future global ecological revolution, directed against even more awesome and entrenched fortresses of power, could be any different?

Once this topic is studied closely, it is hard to imagine a future revolution occurring as a manifestly pacifist event, though precise amounts and types of violence will clearly depend on such factors as the amount of resources (material and ideological) at the disposal of all parties involved. While normal politics can obviously secure reforms peacefully, any serious opposition to existing class and power structures is sure to bring fierce and sustained violent resistance, as it has throughout history. The vision of a peaceful society is surely to be preferred, but that vision is no automatic guide to the strategy needed to achieve it. The problem of violence is thus less one of doctrinaire morality than of political necessity, where violence in the service of change is understood as part of a total arsenal of resources on both sides of the conflict. Since the modern corporate-state rests upon massive coercive power in its multiple realms, its overthrow is highly unlikely within the framework of normal (that is, legitimate and peaceful) politics; more aggressive modes of popular insurgency are crucial to success. The Greens' uncompromising dedication to pacifism (a core principle) overlooks this imperative, inspired as it is by a false reading of earlier historical examples—above all the Civil Rights movement (King) and the mobilization for Indian national independence (Gandhi). Neither example demonstrates what it claims—that nonviolent methods alone can secure radical change.

The Civil Rights example fails on at least three counts. The first, and most obvious, is that the movement never actually set out to fundamentally alter class and power relations; it was reformist to the core, seeking *integration* of blacks into the political system by means of extending citizenship and reforming laws. Resistance came largely from the last vestiges of the old South, from a crumbling power elite already discredited in American society and doomed to vanish. A wide range of "normal" interventions were possible and made sense—legal decisions, voter registration drives, marches, demonstrations, and so forth. The movement scarcely required (or demanded) changes in the basic structures of power, however militant its actions and despite many clashes (some violent) with government officials and racist opponents. Second, the federal government—that is, effective state power—was decisively on the side of the movement, from the passing of laws and policies to the (often flawed) protection of Civil Rights activists in the South and elsewhere. The government in this instance was no hostile behemoth that had to be fought and destroyed. Third, popular struggles to overturn the Southern racist system, going back to the Civil War and earlier, did actually involve untold levels of political violence on all sides. The Jim Crow system was not only enormously coercive and brutal in its repression of blacks, but gave rise to violent opposition as a prelude to postwar Civil Rights gains. Even during the King years, moreover, violence was a common feature of black insurgency, as when Presidents Eisenhower, Kennedy, and Johnson sent federal troops to the South as part of the battle to integrate public institutions. Further, during the 1960s and beyond, newer forms of violent black mobilization (urban riots, Black Panthers) helped create an atmosphere in which Civil Rights goals could be more effectively advanced to the extent ruling elites were forced onto the defensive. In the end, the U.S. federal government was able to exercise legal and institutional power to help facilitate integration.

The Indian example is even less convincing: postwar national independence from Britain supposedly orchestrated by Gandhi in fact proves just the opposite of what is frequently argued. The movement ultimately replaced imperial rulers based in London with an indigenous elite comprising Indian bankers, industrialists, and landlords; there was no revolution, no overturning of class relations, indeed no improvement in the wretched condition of the masses. The chronic ills of Indian society—poverty, oppression, and disease—actually worsened during subsequent decades, reproducing daily violence probably far worse than what would have accompanied mass insurrection and revolutionary change. The partition separating Indian and Pakistan after 1947 brought unbelievable levels of bloody conflict. Gandhi's main allegiance, supposedly to the disenfranchised masses, was really to the Indian national elites who emerged as great beneficiaries of the new power structure, including a Congress Party responsible for new waves of violent repression. In fact the British were already prepared to give up their imperial control of India—Gandhi or no Gandhi—in the wake of their deep exhaustion from World War II and the diminishing returns of colonial rule.

Further, Gandhi's philosophy of nonviolent resistance—grounded in tenets of otherworldly spiritualism, love, good intentions, and self-sacrifice—was never understood as politically transformative, much less revolutionary. Its pacifist tactics were deliberately moderate, restrained, even depoliticizing, designed to channel rather than liberate mass energies. Always close to the Indian elites, Gandhi was stridently hostile to the domestic left that included the labor movement and Marxist parties. Full-scale revolt would have been a mortal threat to an Indian bourgeoisie just beginning to consolidate power on its own economic and ideological footing. In this context, Gandhi's romanticization of traditional social life was perfectly functional to the (renovated) Indian power structure.

Neither the American Civil Rights movement nor the struggle for Indian national independence thus proves what is axiomatically claimed for the superiority of nonviolent resistance. (Indeed the term "resistance" itself suggests a modality quite different from social transformation or revolution.) The same is true for South African efforts to dismantle apartheid, which had involved decades of insurrectionary activity marked by violence on all sides. Even here, sectors of the old power structure—after being stripped of their most brutal racist institutions and practices—were able to reconsolidate elements of political hegemony. The historical actuality is that strict adherence to nonviolent methods sharply reduces strategic options while ceding important terrain to ruling forces always prepared to unleash force when their interests are threatened. No truly revolutionary political formation has ever shown a willingness to do this, as it would have been regarded as suicidal. To be sure, nonviolent resistance, including acts of peaceful disobedience, has often been a prelude to reforms—though always within systemic boundaries. A qualitative shift in class and power relations has always required disruptions, forms of direct action, and popular rebellion aided by varying degrees of violence that, in many cases, was initiated by the ruling interests. The Greens' doctrinaire embrace of pacifism and nonviolent politics, therefore, inevitably gives the elites an upper hand in strategic encounters where maximum flexibility is always crucial. Any future ecological revolution cannot afford to be straightjacketed in this fashion.

Hegemony and Revolution

We have seen how historical models of radical change fall well short of anti-system capabilities, even as the world capitalist order continues full speed along its destructive path. It follows that future transformative struggles will need renewed theoretical and political departures that go beyond two of the most powerful ideological legacies of the twentieth century—liberalism and Marxism. Ecological crisis generates new contradictions and challenges and, with them, the demand for fresh strategic departures. To date, unfortunately, the corporate state has managed to sustain legitimacy in the face of recurring, often severe, crises as well as popular challenges. It is easy enough to attribute capitalist longevity to failed anti-system leadership, bureaucratic corruption, or simple mass immaturity, yet the difficulties run much deeper: the ruling interests have been extremely shrewd at generating popular loyalty across a wide variety of historical conditions and political challenges.

Classical Marxism, unfortunately, failed to enlighten regarding this dilemma, in part because of its exceedingly optimistic belief that the very dynamics of capitalism—expansion of material forces, numerical growth and radicalization of the proletariat, endemic crisis tendencies—would bring revolutionary change. Historical circumstances would drive the working class toward anticapitalist opposition, however slowly or unevenly; exploited and alienated, labor would move toward heightened class consciousness and be driven to create a more rational social order based on human creativity, equality, and socially-useful (rather than commodified) economic activity. Heirs of the Enlightenment, Marx and Engels (followed by their disciples) embraced a psychological rationalism linking human progress with industrial expansion and technological innovation, with capitalism itself laying the groundwork for socialism. In the end, workers—the driving force of history—would logically break with the regime of capitalist hierarchy and discipline as it denied their humanity. Across succeeding decades, however, "objective" historical conditions gave rise to no such anticipated outcomes, as workers only episodically demonstrated zeal for revolutionary action. Advancing economic modernization, as in Europe, generated strong tendencies toward reformism, with the more radical insurgencies receding after the great Italian and Spanish defeats of the 1920s and 1930s. Communist revolutions, on the other hand, were mostly peasant-based, winning power under conditions of heightened nationalist mobilization against foreign rule. The general reality, as both Lenin and Bernstein had already recognized at the turn of the century, was that the industrial proletariat had come under the influence of bourgeois (economistic, reformist) ideology. Lenin and Bernstein were insightful enough: despite harsh exploitation and poverty wrought by capitalism, despite crisis after crisis, the system in most countries retained a degree of legitimacy as workers (and other strata) lent their consent to the power structure. Whatever the national setting, established movements, unions, and parties typically conducted business within the framework of liberal capitalism, as bargaining agents with the political and managerial elites.

By the late 1920s, after bitter defeat of the Italian council movement, the rise of European fascism, and consolidation of Stalinist power in the USSR, hopes for socialist revolution in the West were fading rapidly. Intensified economic crisis did not lead to proletarian consciousness broad or radical enough to overthrow capitalism. Meanwhile, fascist ascendancy was fueled by a resurgent corporatism and nationalism that found fertile terrain first in Italy and then Germany. This was no temporary detour along the road to expanded revolutionary politics, but indicated a far deeper predicament—the long-term capacity of ruling elites to perpetuate their domination through ideological appeals, strengthened by the growth of mass media, education, and kindred influences on popular consciousness. From the viewpoint of both an embryonic "Western Marxism" and Frankfurt School situated in Europe, failed revolution was a matter of historical conditions that would have irreversible consequences for oppositional politics, at least in the West.[20] Confounding initial Marxist expectations, capitalism had been able to solidify its popular support by means of what Gramsci called "ideological hegemony," essentially a reframing of pessimistic conclusions about prospects for sharpening class consciousness arrived at earlier by Bernstein and Lenin. Gramsci's theoretical construct was designed to explain, and locate, the decline of proletarian agency within advanced capitalism—a construct later enjoying intellectual and political resonance in Europe, North America, and beyond. Lenin's epic solution to this conundrum—the Jacobin party—was eventually Gramsci's own (though modified) solution, even as Western Marxism (and the Frankfurt School) rejected vanguardism as yet another source of elite power over the masses.

With its attention to the role of ideas and culture in transformative change, theorists associated with "Western" or "critical" Marxism laid the groundwork for a more complex, multifaceted counter-hegemonic process—that is, a new and broader variant of anti-system politics. In Gramsci's view, advanced capitalism gained legitimation through a vast sweep of ideological appeals—nationalism, religion, preindustrial values, cultural traditions—not always directly traceable to the mode of production or class relations. Strong legitimation meant that ruling interests could rely less on state coercion and more on ideological consensus to sustain domination. Systemic power depended increasingly on mass consent, which early Marxism had relegated to secondary or peripheral status as ideology was understood simply as a ruling-class device to mystify workers. In the 1930s, with the onset of economic depression, Gramsci placed ideology at the center of how ruling-class power is either reproduced or subverted. The broad "ensemble of relations" constitutive of civil society had to be transformed as part of the "war of position" (ideological contestation) leading to "war of movement" (struggle for state power), a process requiring a social bloc of forces with common objectives. Gramsci understood epic shifts in popular consciousness as integral to social forces in motion at specific historical conjunctures—for example, at moments of national resistance against foreign power—rather than simple, unmediated expressions of class conflict.[21] This schema anticipated the rise of later social blocs, including antifascist Resistance movements across Europe during World War II, the postwar Chinese and Yugoslav revolutions, and the new left of the 1960s. Such blocs signify a widening oppositional public space giving substance to an emergent anti-system politics. Whether an ascendant ecological radicalism, or Green politics, currently fits the pattern of Gramsci's social bloc remains to be seen.

Reversing the global crisis necessitates a process of delegitimation building toward a new "conception of life" (Gramsci's phrase), in opposition to the attitudes, beliefs, and myths of "common sense" that buttress corporate-state power. As global warming and related crises worsen, the feeble (largely cosmetic) character of elite solutions could hasten de-legitimation, as old discourses and formulas appear increasingly less rational and workable. This could stimulate oppositional tendencies, at which point the power structure will mobilize every resource at its disposal to prevent further erosion, which today is bound to include deceptive greenwashing, reform palliatives, false electoral promises, media obfuscation, junk science, and perhaps even war making to deflect the public gaze away from pressing domestic problems and toward patriotism—while proceeding full speed ahead with corporate business-as-usual. Although such overtures will do nothing to stave off actual challenges, they can temporarily alleviate legitimacy deficits. Once ideological supports begin to wane, however, as in Eastern Europe and South Africa during the 1980s, radical change can rapidly enter the popular Zeitgeist, as "common sense" loses its resonance among the public and even sectors of the ruling class. What remains to be determined, as always, is the precise ideological *direction* of emergent counter-hegemonic struggles. To the degree the modern crisis opens new space for the expansion of social blocs, it could allow (in Gramscian language) for the merger of oppositional forces where "popular feelings become unified" and bring ideological direction to mass revolt.[22] Such a unifying or "global" dynamic links disparate popular struggles that, left to their own social immediacy, would likely dissolve into a mixture of interest groups, community enclaves, identity politics, and personal alternatives. Social forces in motion, even where popular and militant, do not alone represent historical agencies of change; lacking durable organization, coherent ideology, and workable strategy, they can easily splinter and disaggregate, winding up politically inert. A politically vibrant social bloc depends on an explosive congruence of historical conditions and subjective responses. In Italy, the epic transformation of the Communist Party from a tiny, isolated underground nucleus into a thriving mass organization with two million members during the World War II antifascist Resistance fits this pattern, ascending at the moment of Mussolini's epic defeat. The same logic applies to wartime mobilizations in France, Greece, Czechoslovakia, Yugoslavia, and elsewhere. The ideological cement of these social blocs was overwhelmingly *nationalism*, in this case directed *against* the existing power structure and fueled by a common (multi-class) struggle to evict a despised foreign presence (the Nazis). This same dynamic undergirded such twentieth-century revolutions as the Chinese, Vietnamese, Yugoslav, and Cuban.[23]

It might be argued that such history is not likely to be repeated within industrialized societies today, as nationalist ideology now serves mostly reactionary interests, especially for a superpower like the United States, where militarism and patriotism are ritually invoked in support of imperial objectives and state power. Further, the threat to planetary survival—a manifestly *global* challenge—must be a catalytic mobilizing force behind any future social bloc, which nationalism is destined to oppose. Meanwhile, the corporate-state generates popular anger over distinctly material issues like plutocratic domination, poverty, unemployment, and declining public services likely to be more salient than environmental problems that for most people might appear distant and, in any event, are routinely downplayed in the media. All these challenges are in fact deeply connected, but in everyday public discourse are usually seen in relative isolation as if entirely *disconnected*. The historical experience is that single-issue movements usually morph into mainstream interest groups, which indeed has been the fate of most American environmental organizations. The political meaning of social blocs is that they can transcend this predicament, elevating the "war of position" to a "war of movement" on the institutional (and strategic) terrain.

The present struggle depends on the capacity of oppositional groups and movements, especially in the industrialized North, to create social blocs with enough leverage to confront globalized state power. Could mounting ecological horrors give rise to ideological convergence like that provided by nationalism for earlier revolutions? If so, the global crisis must worsen to where it severely disrupts everyday life, forcing changes in thought and behavior—now a seemingly remote prospect, especially given the popular fixation on economic issues. The ideological terrain to be traversed is daunting, yet hardly unimaginable as the repercussions of global warming and resource shrinkage are more deeply and widely felt. Ecological revolution can make little headway, however, without the erosion of deeply ingrained beliefs and lifestyles: possessive individualism, consumerism, faith in corporate-growth agendas, fetishism of technology, patriotism. A true "greening" of industrial society must involve critical rethinking of Enlightenment values that underpin the whole matrix of domination. Here the spread of ecological consciousness across Europe and North America offers hope—but only if connected to an anti-system political strategy. In the case of the United States, Gramsci argued that the first Anglo-Saxon pioneers had forged a "new level of civilization" untrammeled by preindustrial residues, making possible the virtually limitless expansion of liberal-capitalism. With no feudal restraints, obstacles to capitalist development (and later corporate power) were more easily defeated in the United States than in relatively tradition-bound Europe. Endemic to American society was a public life built around the capitalist economy from which emerged a "new type of human being," a new work process, and thriving materialist culture allowing for strengthened corporate-bureaucratic domination.[24] What Gramsci labeled "Fordism" or "Americanism" further shaped the twentieth-century marriage of liberalism and technological rationality, a force behind the expansion of organized state capitalism in the decades after Gramsci's 1937 death—a phenomenon analyzed by Max Weber and later by C. Wright Mills and Herbert Marcuse.

As centerpieces of a Fordist economy, bureaucracy and technology were integral to both industrialization and the general system of *domination,* constituting dynamic sites of bourgeois hegemony. Gramsci was among the first to theorize the oppressive features of capitalist rationalization rooted in new forms of social and ideological control, the degradation or labor, and blunting of proletarian consciousness as workers took on the characteristics of "trained gorillas." Galvanized by Fordist mass production, the new period would give rise to hyper-rationalized operations such as "McDonaldization" and the Wal-Mart retail revolution, with equally debilitating repercussions for the labor force. Gramsci saw this as not only destructive of workplace autonomy and creativity, but as a resounding threat to political opposition. Beneath the facade of pluralism and diversity, the system would produce nothing so much as hierarchy, subordination, and conformism. The ruling class favored hegemonic discourses and submissive behavioral patterns consistent with maximum corporate freedom. Gramsci's projection of technological rationality as a key factor in mass consciousness paralleled the seminal work of Weber at the time, while anticipating later theories of rationalization by Theodor Adorno and Max Horkheimer (in *Dialectic of Enlightenment*), Herbert Marcuse (in *One-Dimensional Man*), and Harry Braverman (in *Labor and Monopoly Capital*).[25] Insofar as Gramsci's view of capitalist rationalization as a mechanism of ideological hegemony held true, then conventional Marxist expectations of intensifying class conflict leading to proletarian revolution would have to be reframed at a time when the corporate state stabilizes itself through new, more effective forms of popular control. Such theoretical shifts helped explain the failure of proletarian-based revolutions in advanced capitalism, calling attention to a lengthy phase of elite reconsolidation enhanced by technocratic engineering and media manipulation.

Trends identified by Gramsci in the 1930s would become more pervasive and durable with advancing industrialization—hardly surprising given the growth of bureaucracy and technology in the system of production (and domination). Technology would of course come to reshape the entire landscape, from government to manufacturing, the workplace, communications, agriculture, the military, education, and media. The Enlightenment had now achieved its fullest material and ideological expression, refashioning and enlarging modernity for the new period. Not only were social and authority relations further legitimated through technological rationality, so too was the corporate-growth apparatus. The waning of class struggle in advanced capitalism coincided with the surfacing of new contradictions rooted in technocracy and the renewed corporate drive toward mastering nature. It was Marcuse who probably best conceptualized this new dialectic, framing technological rationality as the matrix of "one-dimensionality" marked by mass depoliticization and ritualized (technocratic) discourse integrating the vast majority (including workers) into dominant modes of thought and behavior, noting, "The incessant dynamic of technical progress has become permeated with political content, and the Logos of technics has been made into the Logos of continued servitude."[26]

Marcuse saw technological rationality as central to ideological hegemony in advanced capitalism, narrowing public discourse and subverting political opposition in a society where obedience, routine, and order crowded out critical thinking and social protest. Alternatives to one-dimensional order appeared irrational, beyond the scope of public debate. At the core of Enlightenment thinking, technology functions as both material force and ideology, legitimating class domination in the midst of mounting authoritarianism, inequality, violence, and environmental destruction. What Marcuse called "liberation" meant a thorough "redefinition of needs" grounded in the struggle for new human values that, in the end, would be dependent on the restructuring of both social and natural relations.[27] Because technology is so crucial to the ruling interests, it follows that "qualitative change is conditional upon planning for the whole against these interests, and a free and rational society can emerge only on that basis."[28] Meanwhile, technological rationality gives impetus to more sophisticated forms of propaganda integral to an expanded media culture. Taking Gramsci's analysis of Fordism beyond strictly class relations, Marcuse viewed revolutionary change as engaging the totality of social and authority relations in modern society. Counter-hegemonic politics thus advances through a wider set of contradictions, forcing an epic transformation of human-nature relations.

Marcuse came to understand the connection between nature and freedom, ecology and revolution as the central dialectic of the modern era.[29] Radical politics depends on a flourishing ecological consciousness at a time when systemic conflict is shifting from a strictly class dialectic to generalized opposition against the corporate-state growth regime by a multi-class social bloc, suggesting that "the emancipation of the senses must accompany the emancipation of consciousness, thus involving the *totality* of human existence."[30] Could any phenomenon today be more "totalizing" than the global ecological crisis? Contrary to earlier Marxist assumptions, the entire field of social conflict and historical agency must be reconceptualized to fully engage the new conjuncture of social forces. We have seen how social ecology and left-Green tendencies embrace the notion of diverse popular movements against multiple and overlapping forms of domination—a perspective fully in sync with Marcuse's outlook as well as Gramsci's concept of social bloc. Clearly the diffusion of ecology groups and movements opens space for a distinctive Green *politics,* auguring a new phase of radical agency. As early as 1971 Marcuse could write that "the concrete link between the liberation of man and that of nature has become manifest today in the role which the thrust toward ecological renewal plays in the radical movement."[31] The central question at this point is whether, and to what extent, an ecological politics can flourish as a counter-hegemonic opposition in the modern setting.

Similar ecological departures were laid out by Bookchin, Bahro, and others within left-Green politics during the 1980s, Bookchin writing, "The success of the revolutionary project must now rest on the emergence of a general human interest that cuts across the particularistic interests of class, nationality, ethnicity, and gender."[32] Such ecological radicalism depends on the politicization of social movements, the centrality of natural relations, a focus on multiple forms of domination, and transcendence of postmodern identity agendas. For Bookchin, the turning point of radical change is that "no general interest of this kind can be achieved by the particular means that marked earlier revolutionary movements."[33] As noted, the great mobilizing force of nationalism in twentieth-century revolutions will thus be counterproductive to any future ecological politics. A new radicalism must address not only deepening global crisis—that is, possible descent into barbarism—but a corporate-state fortress that is oligarchic, ruthless, and largely immune to genuine reform in an era when political decision making is lodged in the hands of fewer and fewer elites. A post-liberal, post-Marxist transformative politics is urgently needed in the face of unprecedented social and ecological demands: sustainable modes of production, work, and consumption, a political system freed of moneyed interests, equitable health care, an independent media, diffusion of green technologies, localized agriculture, a post-carbon economy, demilitarization, and urban revitalization—all infused with democratic governance.

Conclusion: A Green Politics?

The overriding problem for early twenty-first-century struggles to create an alternative to crisis-ridden world capitalism revolves around bringing local movements into a political matrix with enough ideological and organizational force to confront the global challenge. At present thousands of community groups and grassroots movements, including well-funded NGOs and kindred activities, exist across the globe, surely a launching point for any ecological revolution. Reflecting on the fragile state of "grassroots globalism," Tom Mertes refers to "an ongoing series of alliances and coalitions whose convergences remain contingent," which, despite its enormous potential, is essentially pre-political.[1] Many such organizations—hundreds by conservative estimate—are environmental, though only a few are directed against corporate power. The extent to which such widely dispersed movements can become politicized will go far in determining the human capacity to stave off global collapse. Given the uniformly profit-driven logic of corporations, basic reform of world capitalism seems rather delusional since the system would still run counter to ecological values and priorities. One problem is that existing nation-states operate primarily as tools of global capital, while established political parties remain embedded within governmental networks and their anachronistic state-centered nationalist ideologies. The American power structure, with its megacorporations, Wall Street, permanent war economy, and security-state, constitutes an especially imposing obstacle, a vast barrier to revolutionary strategy for a counter-hegemonic, ecological politics.

If grassroots movements represent a natural point of departure, as in the case of the European Greens, the worldwide diffusion and integration of such movements will require an urgent (and rapid) shift toward ecological consciousness—a process now well advanced in some industrialized and even lesser-developed societies. Much depends on the extent of material and social disruptions resulting from the crisis, as such disruptions are sure to force a rethinking of everyday norms and lifestyles. A spreading ecological Zeitgeist means a popular turn toward greater social engagement, sustainable lifestyles, and moderated private consumerism like that embraced by numerous environmental groups and movements. Old practices must be understood as ethically, or at least materially, problematic A consciousness shift of this magnitude depends on the expansion of a critical educational process with ecological thinking at its center, what Peter McLaren calls "ecosocialist critical pedagogies," which could find fertile terrain in schools, universities, and communities.[2] The recent explosion of environmentally themed books, magazines, articles, films, documentaries, and lecture materials has given this ideological ferment a powerful boost. At the same time, space for critical intellectual discourse and consciousness renewal must be forged largely outside the realm of corporate, governmental, and media power.

During the twentieth century such oppositional energies were usually channeled through local movements, trade unions, workers' councils, and popular assemblies, achieving political articulation through the formation of socialist and Communist parties that, in many cases, had been able to build strongholds of union, municipal, and governmental power. In one form or another, leftist insurgency had to face the inevitable question of *political strategy*, which, unfortunately, has been largely sidestepped or dismissed within the contemporary setting. While the political imperative remains, decline of those earlier traditions poses ideological barriers to the much-needed translations of environmental struggles into mature organization, leadership, direction, and methods. In many quarters "politics" is dismissed as a toxic source of corruption, impasse, and futility. For many the Soviet collapse revealed nothing so much as the bankruptcy of state power as a conduit of progressive change, returning to prominence the dynamic role of "civil society." The problem is that while social movements continue to flourish across the landscape, their *strategic* potential remains in doubt at a time when the global system has moved into a new phase of development.

While Green parties have built political leverage in many countries, above all in Europe, their trajectory in the United States so far offers few grounds for optimism. In a context where grassroots-centered agendas and ecological discourse have permeated the culture, liberal environmentalism—discussed at length in Chapter 3—currently predominates. A leading example is the Post-Carbon Institute, a progressive center calling for radical alternatives to the broken American economic and political paradigms, including cutbacks in private consumption, a phasing out of fossil fuels, and a turn toward local economic arrangements, but offering little clarity about general objectives or political methods for advancing them. Referring to general Institute ambitions, David Orr urges adoption of a "new political framework," writing, "Our situation calls for the transformation of governance and policies in ways that are somewhat comparable to that in U.S. history between the years 1776 and 1800".[3] Orr points toward an institutional framework in which carbon emissions can be stabilized and ultimately reduced, questioning a system that favors corporate profits over public goods. Yet nowhere is there mention of political strategy—or the role of movements and parties needed to translate such a strategy into durable social outcomes. Rather, quite strangely, Orr calls attention to the role of presidential authority in creating "transformational leadership" to lay the groundwork for a post-carbon society.[4] Existing structures, including those presumably viewed as part of the "broken paradigm," would from all indications remain fully intact.

Where Greens have become a mainstream but still marginalized political force, as in the United States, radical change depends on something akin to an ecosocialist trend—inside or outside the party. McLaren's call for "ecosocialist critical pedagogies" can be viewed as the necessary first step in the process. While American society has witnessed the proliferation of such groups and movements at different moments of its history, the distance from socialist or Marxist politics has actually widened since the 1970s, and emphatically following the Soviet collapse. Could the socialist legacy somehow be resurrected and brought into the Green orbit as the global crisis intensifies? Or might the "socialist" dimension of ecosocialist politics create new ideological barriers to any potential ecological radicalism? John Nichols, for one, believes that the socialist tradition remains to some degree a vital part of American political life, even as it continues to be demonized by the media and political establishments. In *The 'S' Word*, Nichols argues that some 30 percent of Americans have a positive view of socialism, understood generically as a democratic, egalitarian alternative to dysfunctional state-capitalism, which such a large percentage of Americans has come to loathe.[5] Nichols identifies a vigorous socialist tradition that shares much in common with the progressive side of liberalism—a tradition badly needing to be revived in the midst of a worsening cycle of economic, ecological, and military challenges. At the same time, Nichols scarcely refers to the environmental crisis or the urgency of reaching developmental sustainability: his version of socialism—a more democratized and public-centered liberalism—shares little with the ecosocialism championed by McLaren, Foster, Chris Williams, and this author.

In their call for an ecosocialist revolution both John Bellamy Foster and Chris Williams demand a jettisoning of the world capitalist system in favor of a radically egalitarian and democratic society (Foster) made possible by a "total remodeling of the world on all levels" (Williams). Both are convinced that, in the absence of a global ecological revolution, planetary catastrophe is imminent. According to Foster, the truly planetary crisis we are now caught up in demands a world uprising "transgressing the boundaries between humanity and the planet."[6] Environmental balance requires a transition from capitalism to socialism, which, of course, cannot occur without a political-strategic merger of oppositional forces similar to what defined twentieth-century parties and states. Williams argues for "gigantic systemic change" in which "we collectively and democratically make all decisions based on human need, not corporate profits".[7] Such ecosocialist visions are appealing, to be sure, but like many contemporary arguments for radical change wind up avoiding the thornier questions of political strategy. When it comes to decisive issues of organization, ideology, and methods we are simply left to the realm of conjecture. Will the global ecological revolution be the province of a Leninist vanguard, a social-democratic parliamentarism, an anarchist-based insurrectionary process, a Green radicalism—or what? It is tempting to defer crucial political debates into a vague future: after all, the ideological vision remains strong while the "crisis" can be expected to force the issue as a matter of unfolding consensus. Yet the overpowering urgency posed by the global crisis unfortunately permits no such temporizing.

It has been left to Green party theorists and activists, essentially by default, to articulate the broad contours of a radical ecological strategy—a strategy that over three decades has had its share of tactical conflicts and adaptations to meet new conditions, not to mention outright failures. Joel Kovel, for example, frames a Green politics that envisions a process of party building from the ground up, transcending the historic alternatives of Leninism, social democracy, and anarchism as politically inconsistent with an ecological radicalism geared to the struggle for state power. Embracing consciousness transformation and grassroots insurgency as starting points of change, Kovel looks to a political structure imparting first national and then global substance to ecological politics. The process moves ever upward, toward formation of a social bloc and a political party firmly grounded in local movements. Thus, "The general model of ecosocialist development is to foster the activity potentials of ensembles in order to draw together those points into even more dynamic bodies."[8] This essentially Gramscian politics is congruent with the kind of Green radicalism advanced by ecosocialists in Europe but which, unfortunately, has all too often fallen victim to the logic of deradicalization in a context where the Realos, in pursuit of a strict electoral strategy, became dominant. Kovel nonetheless retains faith in a Green radicalism that, while attentive to electoral and parliamentary politics, can build on hundreds, perhaps thousands, of local struggles, community activities, and social movements that are sure to thrive as the global crisis worsens and forces ordinary people away from present institutional arrangements and social-life patterns.[9]

Postscript: Ecology and Population

With the deepening global crisis and the increasing likelihood of planetary disaster, the impact of unsustainable human population growth must be addressed, as the world total is projected to reach 9.1 billion by 2050 and easily more than ten billion by 2100. Could the burdensome demands of such an enlarged human footprint on the earth's carrying capacity somehow *not* be a crucial factor in world ecological (also economic) calculations? Chris Hedges states what must be starkly obvious: "All measures to thwart the degradation and destruction of our ecosystem will be useless if we do not cut population growth."[1] After all, population levels are a great multiplier of global amounts of production and consumption, energy utilization, and general resource demand. In a world of terribly strained natural resources, swollen urban centers, climate change, and sharpened geopolitical conflict the issue of overpopulation surely confronts humanity as a key causal factor in the modern predicament. If this is true, then *population control* will be strategically central to all challenges ahead: global warming, shrinking cropland, declining forests, food shortages, increased poverty, health problems, air and water pollution, and resource shortages including water, timber, and oil. Future transition to an ecological model of development, based in a post-carbon economy, will mean stricter limits to both industrial and population growth—limits even now resisted by the dominant interests. As scientific discourse and environmental pressures both pose the imperative of creating an ecological politics, the total number of human beings dwelling on planet earth must inevitably figure in the strategic equation. Strangely, however, the population question is generally ignored or downplayed in public discourse, even among most environmentalists: the prevailing view is that the economy and the population should both be allowed, in fact, encouraged, to grow without limits, as vital indicators of human progress. Of course, the extreme right has its own emotionally charged (religious and ideological) pretext for avoiding debate on the population question. A few liberal environmentalists (Lester Brown, Paul and Anne Ehrlich) who insist on a linkage between the ecological crisis and population growth are ritually attacked as misanthropic "Malthusians," even racists, ever fearful of teeming Third World masses. Mainstream liberalism, on the other hand, wants little to do with such a "controversial" topic. Empirical discussion of population issues—indeed anything hinting at limits of growth—seems taboo in an increasingly conservative ideological climate.

A May 2011 UN report on population growth concludes from newly calculated demographic trends that world population could reach 10.1 billion people by the end of the century—a nearly 50 percent increase beyond seven billion in 2011. Contrary to what is argued by overpopulation deniers, this report surpasses previous expectations. The highest growth rates are anticipated for the poorest countries, meaning regions where environmental sustainability is most difficult to achieve. In Africa, total population is expected to rise threefold, reaching nearly 3.5 billion people on a continent already struggling to feed its people. Projections for Nigeria alone are staggering: growth from 162 million to 730 million people by 2100. While the European continent is predicted to stabilize as fertility levels decline, the report indicates that U.S. population will expand from 311 million to 478 million by the end of the century.[2] While such demographic projections can obviously be arrested or reversed, this will require far-reaching population control efforts, which so far have been dismissed across the political spectrum, especially in the United States.

Left-wing critics have been especially hostile to population discourse—a case in point being Chris Williams in his *Ecology and Socialism* (2011). Williams attacks Brown and the Ehrlichs as "resource-depletion doomsayers" whose claims regarding the "population explosion" merely reinforce the ideological needs of the system; they are misleading, even incendiary.[3] Any diffusion of the population growth myth, and with it the false notion of finite earthly resources, belongs to the "ideological armory of capitalism" as it deflects attention from the real origins of the ecological crisis—above all, the destructive consequences of the world economy.[4] Williams' argument revolves around three familiar claims: the "doomsayers" are Malthusian, they fail to recognize that world population has a slowing "growth rate," and, most crucially, they ignore the dynamic factor of "social relations" in shaping the use and distribution of global resources.

The first point—Malthusian ideological bias—is just as distressingly predictable as it is totally irrelevant. Williams follows the formulaic path of saddling overpopulation theorists with Malthusian "class prejudice," an ideological ploy seemingly buttressed by the (well-established) conclusion that Malthus' own projections of geometric population increase have been falsified. Since Malthus was shown to exhibit a profound contempt for the poor masses, this clearly must have been a motivation behind his obsession with population control in a world that, in the late eighteenth century, had less than one billion people. Williams cites several passages from Marx, whose scathing criticisms of Malthus for ignoring the role of capitalism are well known—and indeed Marx was perfectly correct in his attacks. The problem is that Malthus' outmoded views of social development have nothing to do with the contemporary ecological situation, which bears little relation to Malthusian ideas or the world he inhabited. Whether or not population expansion is geometrical or simply incremental is of arcane interest on a planet that cannot presently sustain a level of seven billion humans much less a projected total of ten billion or more. Planetary catastrophe could scarcely be imagined two centuries ago. The era of Malthus (and indeed Marx) has so little in common with the present-day global crisis that we are best advised to simply forget about Malthus and those remote debates. Further, none of the targeted critics (Brown, the Ehrlichs, et al.) claim any specific lineage with Malthus or rely heavily on his work—nor, more importantly, do they exhibit the extreme class prejudice of Malthus and his earlier disciples. The Malthusian bogeyman is an ideological detour that should finally be laid to rest.

Arriving closer to the issue at hand, Williams argues that population growth today matters little to the larger ecological challenge; it is grossly overstated, a distraction from the more urgent concerns of corporate power and world poverty. Williams recognizes the problem of global sustainability but contends that it has nothing to do with population levels. He refers to the logic of a "falling rate of population increase": although world population continues to rise, its growth rate actually peaked in the 1960s and it continues to grow only slightly as the human community ages and reaches declining fertility rates. For a variety of reasons, moreover, heightened economic development functions to limit family size, as it has done in regions such as Europe, Williams noting, "This reduction in growth rate undermines the presumption that human population has reached the earth's carrying capacity."[5] He states that, according to demographic projections, by 2030 the world will comfortably be able to feed at least eight billion people. Leaving aside here the problem of Williams' exclusive attention to the question of food shortages, his conclusion resonates little for a planet already overcome by global crisis marked by climate change, the onset of peak oil, diminished water supplies, soil erosion, vanishing arable land, and deforestation. Even without future population *growth*, the world will be a drastically altered habitat with threats likely to be unmanageable in 30 or 40 years, if not sooner. For a Marxist, Williams' critique is strangely undialectical and lacking in historical perspective.

Reliance on a "falling rate" of population increase to show that the earth's carrying capacity is not being threatened ends up irrelevant, since what ultimately matters is the *absolute* number of humans on the planet and their consumption levels, relative to the general resource base and natural habitat resiliency. A reduction in growth *rate*, moreover, is naturally expected given larger population totals with each passing year; what matters is absolute numbers over time and space. At present levels even small increments of population growth will be hugely consequential. We know that global population continues to expand at roughly 80 million yearly, slowing minimally, if at all, until at least 2050. Mention of lowered fertility rates in Europe is likewise meaningless, as Europe contains less than 10 percent of the world's inhabitants. With one billion people on the planet in 1830, the numbers rose to two billion by 1930, five billion by 1990, and nearly seven billion by 2011, with estimates for 2050 at 9.1 billion, as mentioned. Not even Williams insists that such large numbers (whatever the "rate of increase") will be unproblematic for the future health of ecosystems. Even leaving aside crucial food issues, how could it be rationally argued that billions more people will not exert unsustainable pressures on natural resources and the fragile infrastructures of already congested world cities—not to mention the impact of climate change? Diatribes against "Malthusianism," here as elsewhere, contribute nothing to our understanding of population pressures and how they fuel the global crisis.

Population growth in India is especially illustrative of the way natural habitats are being stretched beyond tolerable limits. According to census data released in June 2011, the Indian population has expanded to 1.2 billion, an increase of 181 million in just a single decade. Reports further indicate that population size will more than *double* over the next 50 years, despite a decline in *growth rate* to 17.6 percent relative to the previous decade. The population of just two Indian states now exceeds that of the entire United States, with demands on arable land, water, and other resources—as well as growing urban infrastructures—destined to far exceed the national capacity to satisfy such demands. While poverty increases and already compromised ecosystems deteriorate, economic "development" in India is programmed to benefit a tiny stratum of the most wealthy and powerful. Meanwhile, with nearly 30 percent of the Indian population illiterate, the Congress government allocates less to education each year, thus undermining a crucial linchpin of efforts to curb population growth and maintain a semblance of ecological balance.[6]

The problem of absolute numbers should not be confused with the issue of *population density* with which Williams and others seem needlessly distracted. Density is one factor among many, but in itself is hardly central to prospects for environmental sustainability. Dense populations, as in many Northern urban areas, can be more or less in sync with their natural surroundings. The real question here involves the distribution of inhabitants relative to the local (and global) carrying capacity of ecosystems. The moment at which overpopulation is reached is that of ecological imbalance—that is, when resources are being depleted at a faster rate than they can be renewed. Thus, while even the most developed urban centers might appear sustainable when viewed in isolation, they typically achieve this only by importing (or exploiting) vast amounts of resources (oil, water, food, etc.) and labor from elsewhere. Much depends, therefore, on how specific populations utilize resources in the global context, which transcends the issue of density. Trends toward unsustainability—mostly the result of excessively high-energy production and consumption in the wealthier nations—cannot be attributed to density alone—nor do the Ehrlichs and Brown, for example, make such claims. Contrary to their supposed Malthusian contempt for the poor masses, they lay primary blame for overpopulation on the rich countries of Europe and North America. In one article the Ehrlichs go to extremes to point out that, while Holland can effortlessly support a population of 1,000 persons per square mile, because it can import massive amounts of oil, fresh water, foodstuffs, and other critical resources to the extent most other parts of the world cannot.[7] Currently, a relatively small concentration of highly affluent people accounts for some two-thirds of environmental destruction, including carbon emissions, mainly owing to its lopsided utilization of fossil-fuels. The unfairly criticized Ehrlichs are clear on this point: "The key to understanding overpopulation is not population density but the numbers of people in an area relative to its resources and the capacity of the environment to sustain human activities; that is, to the area's carrying capacity."[8] While most nations of the world can be viewed as overpopulated to some degree, it is the industrialized North that continues to be the worst offender, the Ehrlichs noting that "almost all the rich nations are overpopulated because they are rapidly drawing down stocks of resources around the world."[9] Such passages scarcely reveal the ideological temperament of Malthusian racist contempt for the world's poor masses. On the contrary, the Ehrlichs emphasize a central dynamic of the population equation—namely, that advanced countries manage to build their own wealth, while maintaining the appearance of ecological balance, by importing resources (above all energy supplies) from outside their own borders and by exploiting myriad labor and material needs to reproduce their prosperous economies.

In holding to the "shibboleth of absolute overpopulation," Williams builds his argument around the destructive consequences of world capitalism, of a neoliberal regime that engenders poverty, social inequality, lopsided resource distribution, and ecological degradation. The problem is not too many people but rather "... historically how many human beings the earth can support depends primarily on the level of productivity of the existing population and the social relations within which they are embedded."[10] The reason food problems have surfaced in recent years has everything to do with the specific nature of *social relations* and little do with resource availability. Williams' argument contains a certain plausibility: as I have argued throughout this book, untrammeled corporate power is a decisive factor in the worsening global crisis. The question arises, however, as to precisely how much difference a transformed system allowing for more egalitarian resource allocation would make given the general constellation of factors at work—climate change, shrinking arable land, declining water tables, peak oil, deforestation, and so forth. Even radically egalitarian arrangements, beneficial as they might be, cannot save the world natural habitat from its downward spiral once the sheer human burdens placed on infrastructure, resources, and climate are taken into account. Further, Williams' argument presumes infinite amounts of crucial resources into the distant future, despite research pointing toward heightened problems concerning arable land, soil quality, water supplies, pollution, and energy sources. A dramatic shift in class and power relations would help alleviate the crisis, but not so much that population levels would be irrelevant to the broader ecological calculus. There is little to suggest, however, that fundamental changes in global "social relations" are on the near horizon; existing patterns of production and consumption, however attenuated by way of crisis or reforms, are unfortunately likely to continue for decades, if not longer. Anti-system groups, movements, and parties, as we have seen, remain fragmented and weak in the second decade of the twenty-first century. Even allowing for major policy interventions, a confluence of developmental trends, depleting resources, elevated pollution levels, and population increase is certain to aggravate already imposing ecological challenges ahead. Leftist critics of the overpopulation discourse seem trapped in an ideological denial, apparently convinced that the mounting global crisis will disappear once a more rational economic model is adopted, with skepticism relegated to the minds of Malthusian "doomsayers."

The limits of Williams' perspective run deeper yet: his egalitarian alternative to the world capitalist system, while attentive to ecological priorities, follows a flawed Enlightenment rationality that identifies human progress with endless technological and industrial growth. Capitalist modernization, with all its contradictions and dysfunctions, would be supplanted by an alternative modernity rooted in the same high levels of production, consumption, and resource utilization. Williams nowhere lays out a distinctly ecological model geared to limits of growth. The question is whether the Enlightenment paradigm, firmly embedded in the domination of nature, could ever be sustainable in a world where many developing nations are striving to reach the affluent lifestyles of wealthier Northern countries. Without significant curbs on both economic and population growth—North and South—the Enlightenment scenario (full-speed growth under a socialist regimen) turns out to be just another recipe for ecological disaster, with the natural habitat (and thus all societies) sacrificed for the (illusory) virtues of modernity and progress. Exploitation of nature under any ideological banner or social order ensures perpetual human onslaught on global ecosystems—and likewise the reproduction of *other* modes of domination (class, gender, racial, bureaucratic).

When it comes to population, leftist critics usually arrive at static, ahistorical views of the future: existing conditions are considered somehow immune to worsening trends even with the addition of two or three (perhaps several) billion people exerting new pressures on the planet's carrying capacity. Some, like Williams, take comfort in the fallacy of a "falling rate of growth" or in prospects for a rather quick, radical transformation of "social relations"—but neither of these discourses is currently very compelling. The specter (indeed reality) of severe overpopulation points toward the incapacity of societies to chart sustainable development, threatening the quality of life for *everyone* on the planet, not to mention survival itself. People will need, and surely demand, adequate public goods and services in the form of housing, transportation, health care, education, and sanitation. Reflecting on *contemporary* urban crises across the world, population levels are indisputably a central factor in the downward ecological trajectory. Lacking significant population curbs, even the most ambitious reforms—strong industrial regulations, flourishing alternative technologies, green transportation systems, sustainable agriculture, et cetera—will at best end up only marginally effective. Hedges is correct in observing that all initiatives to counter ecological disaster will be negated in the absence of efforts to reduce population growth, which, by 2050, could easily *double* the population levels of 1980.

The problem of shrinking worldwide food resources relative to demand will probably mark the tipping point for future ecological collapse. A deadly convergence of factors—dwindling arable land, shrinking water supplies, air and water pollution, global warming, and population growth—can only aggravate the crisis in coming years and decades. The major trends are in fact irreversible so long as existing patterns of production, consumption, and resource use persist into future decades. The planet currently loses 1 percent of its arable land yearly, a trend likely to accelerate as damages from climate change become more visible. Per capita cropland for grains, vegetables, and fruits is now diminishing steadily, especially in Russia, China, and India (with roughly 40 percent of the global population), the result of soil erosion, urbanization, desertification, declining water tables, and ongoing transfer of land to livestock grazing and other types of animal farming. Few nations have been able to escape this downward trend, as more countries each year look to import grains and other foodstuffs. World grain production expanded modestly between 1950 and the early 1980s, but with loss of arable land output has virtually stalled since then, the problem compounded by vast increases in the percentage of grains (mostly corn and wheat) fed to animals and more recent initiatives to convert fuel out of grains. A pressing question is how long such nations as China, India, and Brazil can manage high levels of development, accompanied by sprawling urbanization, before excessive loss of cropland turns into a food catastrophe—with obvious global repercussions. And this does not take into account a likely future of droughts, floods, and dust storms produced by global warming.

Williams' glossing over of global warming in relation to overpopulation is most evident on water issues. McKibben observes that "We're every day less the oasis and more the desert. The world hasn't ended, but the world as we know it has—even if we don't quite know it yet."[11] Of all natural resources, water is surely most indispensable since, as the basis of life, human survival fully depends on it and there is no substitute. Further, climate change already deeply impacts water availability in many ways, as planetary reserves of fresh water are depleted through droughts, floods, and melting glaciers that severely impair human capacity to access and control such resources across the globe. Water tables have been eroding for some time, partly because of excessive aquifer plundering for agricultural production; many lakes, rivers, and streams are drying up or have become polluted. Sandra Postel writes that future water challenges will be unprecedented, aggravated by massive increases in human demands.[12] Between 1950 and 2009, as world population grew from 2.5 billion to nearly seven billion, global renewable water supplies declined per person by an astonishing 63 percent, with decline expected to reach another 15 percent by 2025. China is one country already in the midst of a great water crisis: with about 20 percent of the world's people, it possesses just 7 percent of total water resources, and those are sharply declining. Around the earth, more than one billion humans lack safe drinking water, exacerbating already worsening health problems. Water shortages, moreover, put world food supplies increasingly at risk, and this is no mere doomsaying. At present, nearly 70 percent of water is used for agriculture, compared with 20 percent for industry and only 10 percent for human consumption. Not surprisingly, Americans use water at roughly *double* the world average, in part owing to a meat-centered diet and a fast-food culture that has no precedent.[13] The global crisis imposes new pressures on water availability at a time when population growth and increased meat production drastically raise demands on this irreplaceable resource, and quick remedies are nowhere in sight: huge projects (dams, reservoirs, desalinization plants) are expensive, energy draining, and make up a relatively small percentage of the deficit. The world now operates some 15,000 energy-consuming desalinization facilities but these satisfy far less than 1 percent of all human consumption, with little change in sight.

The widening gulf between human needs and agricultural output revolves most imminently around the problem of water shortages, which forces a critical assessment of the (usually ignored) issue of meat production—a problem that Williams, like most leftist critics, ultimately sidesteps, perhaps fearful of addressing long-established dietary (i.e., "personal") habits shared by the vast majority of ordinary people. When it comes to meat, American environmentalists of all political outlooks remain fully in denial, a stunning lapse in a world where animal farming has become the single most destructive cause of resource depletion, pollution, and climate change. As water tables shrink, arable land erodes, soil degrades, overgrazing increases, and deforestation worsens, meat production consumes five times the water of plant-based foods and roughly 40 percent of the world grain harvest.[14] With arable land vanishing most noticeably in large food-producing nations like Russia, China, India, and the United States, fully 1,000 tons of water are required to grow one ton of grain. In the United States, which accounts for 23 percent of worldwide meat and dairy output, the average diet uses a total of 5,000 liters of water daily, wildly unsustainable when projected across the world.

World meat intake rose from 44 million tons in 1950 to 290 million tons in 2008—an average per person increase of 86 pounds—with no decline in this trend presently visible. As nations accelerate their economic growth, meat consumption likewise expands as animal-based diets are universally cherished as signs of "prosperity" or "progress." Most lesser-developed societies look to the American "norm" for dietary aspirations on the (grossly mistaken) premise that such practices are sustainable and healthier, the measure of social and individual robustness. But these consumption patterns can only further exhaust water supplies, destroy land, hasten oil depletion, worsen food shortages, aggravate health problems, and intensify the dynamics of global warming. In the United States, as many aquifers run dangerously low and wells dry up, the water needed to produce one pound of beef is 5, 200 gallons, compared with just 23 gallons for one pound of lettuce, 24 gallons for one pound of potatoes, 25 gallons for one pound of wheat, and 49 gallons for one pound of apples. Indeed, fully *half* of all water used in the United States is devoted, directly or indirectly, to livestock.[15] A meat-based agriculture further depletes the planetary store of soil and water, and increases air pollution from pesticides, fertilizers, and other toxic runoffs. Quite obviously, more global inhabitants, more "development," and more agriculture devoted to meat production spells ecological catastrophe—hastened, of course, when the horrendous impact of the fast-food industry is taken into account.[16]

There is more to this distressing story: old ecosystems, including precious rainforests, are being rapidly destroyed to clear vast regions for meat and dairy agriculture. Cattle grazing in such new zones exacerbates the carbon buildup, since about 25 percent of corrosive methane emissions comes from livestock while the world is robbed of oxygen supplies needed to counter the buildup. In fact meat production devours vast fossil-fuel resources from beginning to end, in the form of pesticides, fertilizers, transport, processing, and food delivery. Prevailing corporate agricultural systems not only hasten oil depletion but in turn will be hit hard by that looming eventuality. While production of one calorie of beef now requires 54 calories of fossil fuels, the amount for most grains is roughly three calories, or just 5 percent of the energy required for meat consumption. As industrializing nations expand animal-based farming, grain stores available to humans are bound to decline sharply, bringing the world ever closer to food disaster. Americans consume 800 kilograms of grain per capita yearly (four times that consumed in India), but only 100 kilograms is eaten directly; the rest goes to livestock and poultry.[17] People in the United States mistakenly believe that living high on the food chain elicits great health benefits, an outlook shared by growing populations in such countries as China, Brazil, and Mexico looking to emulate the same "prosperous" lifestyles. Meanwhile, dozens of nations are now scrambling to secure diminishing grain resources in support of growing meat and other demands, but how long such desperate maneuvers can stave off food crises in an era of declining per capita world grain outputs is problematic. The United States currently furnishes no less than half of all international grain exports, but that will no longer be sustainable at a time of skyrocketing global meat consumption. Meanwhile, increased conversion of grain into fuel for autos, begun on a large scale in 2005, is sure to exacerbate looming food shortages.

On a planet where more than a billion people are drastically underfed, at least two billion—habituated to rich, costly, and carbon-intensive animal-based diets—are wastefully *overfed*, with all the attendant health and environmental problems. As food prices everywhere predictably rise, moreover, the wealthy will still have plenty to eat while the increasing numbers of poor will be condemned to severe malnutrition, even starvation as in the case of Somalia and some other African nations in 2011. No wistful projections of a more rational world system can delay or short-circuit this horrific state of affairs. As the earth veers toward ecological collapse, familiar reassurances by elite opinion (sadly echoed by many leftists) that continuous economic and population growth is perfectly manageable take on a gruesomely dystopic character. Williams' glib contention that food output has always matched population growth, though not remotely true, has obviously less relevance to *future* projections. The question today is whether food availability can satisfy the needs of nine or ten billion people in a period of accentuated climate change, depleted oil, shrinking water tables, worsening per capita agricultural output, and sharp increases in meat consumption. This urgent question—stonewalled by most environmentalists—would have to be taken up even if Robert Malthus had never written a single word.

While overpopulation critics like Brown and the Ehrlichs have been pilloried as neo-Malthusian ideologues, they are fully justified in calling attention to the epochal challenge of sustainability in a world already saddled with resource and food shortages. The aforementioned 2011 UN report, which revises population growth estimates substantially upward, follows the same line of thinking. The issue of overburdened carrying capacity cannot be detached from the tens of millions of inhabitants making (ever-larger) demands on the global ecosystem each year—and it is the absolute population level relative to available resources, not rate of increase or density patterns, that is ultimately decisive. Continued population growth under *any* political condition is guaranteed to fuel environmental disaster. Sustainability demands that worldwide resource use be replenishable, minimally harmful to the environment, and sufficient to meet the basic needs of humans everywhere. Beyond that, it ought to support a comfortable quality of life taking into account a world of finite and, in many cases, threatened resources. At one point in his dismissal of overpopulation arguments Williams does concede a telling point, noting, "Obviously, population is not a completely irrelevant consideration when it comes to food provision and there is an obvious difference in resource use when there are more people."[18] This concession, however, nowhere actually figures in Williams' general assault on the ideological disease of "Malthusianism."

Measures for population control will have to be at the center of any future ecological politics if, as Hedges argues, all other initiatives—no matter how ambitious—are not to be squandered. Brown is often attacked for his supposed extremism on this issue, yet his proposals are relatively mild—including such measures as universal education and improved family planning designed to arrest population levels at roughly eight billion by midcentury.[19] He adds an important caveat that the planet, given present U.S. consumption rates, could in the final analysis support no more than 2.5 billion people. David and Marcia Pimentel offer a far more realistic assessment, concluding that a global population of no more than two billion should ideally be reached to ensure long-term ecological balance, a target requiring at least several decades to achieve.[20] Along with radically transformed patterns of economic life, including a fundamental shift in agriculture tied to a stronger reliance on plant-based foods, a number of far-reaching measures will have to be adopted: effective family planning programs, wide availability of birth control information and services, enhanced literacy and general education, and possibly taxation and other incentives favoring a one-child-per-family policy as in China. A naive reliance on "development" to render population growth more sustainable will have little if any impact on mounting resource and food crises or the larger threat of ecological collapse. Unfortunately, such collapse is the inescapable fate of societies that ignore resource limits, that proceed with business-as-usual as if resources were infinite, that drastically overreach in their grandiose economic, political, and military ambitions. There is no avoiding the contemporary ecological imperative: the post-carbon transition to a new system will have to accept imposing limits to both industrial and population growth.

Notes

Series Editor Preface

1. Donella H. Meadows, Dennis L. Meadows, Jorgen Randers, and William W. Behrens III, *The Limits to Growth: A Report to the Club of Rome's Project on the Predicament of Mankind*, 2nd ed. (New York: New American Library, 1975).
2. See my, *The Death of Industrial Civilization: The Limits to Economic Growth and the Repoliticization of Advanced Industrial Society* (Albany: State University of New York Press, 1990), p. 161.
3. See Plato, *The Republic*, 471c; translated by Allan Bloom (New York: Basic Books, 1968), p. 151.
4. Plato, *Ibid.*, 472a; p. 152 of Bloom's edition, *Ibid*. For this information on the Greek view of the third wave, see Bloom, p. 460, note 35.
5. See Plato, *Ibid.*, 369c; Bloom edition, *Ibid.*, p. 46.
6. See, David Harvey, *A Brief History of Neoliberalism* (New York: Oxford University Press, 2007).
7. Space limitations do not permit a discussion of this prescription by Boggs of the Green worldview as the needed political means for ecological revolution.

Chapter 1

1. On the Intergovernmental Panel on Climate Change scientific reports, see www.ipcc.
2. See this U.S. Department of Energy (DOE) report at Mother Earth Network (www.mnn.com).
3. *Los Angeles Times* (November 6, 2011).
4. www.unfcc.int./2011
5. See the *Washington Post* (May 5, 2011), www.WashingtonPost.com/2011
6. See Lester R. Brown, *Plan B 4.0: Mobilizing to Save Civilization* (New York: W. W. Norton and Co., 2009), p. 31.
7. Michael T. Klare, *Resource Wars: The New Landscape of Global Conflict* (New York: Henry Holt and Co., 2001), p. 7.
8. See Klare, *Resource Wars*, chs. 1–3.
9. See Chris Hedges, *Death of the Liberal Class* (New York: Nation Books, 2010), ch. 3, and Tariq Ali, *The Obama Syndrome* (London: Verso, 2010), ch. 3.

10. Cornelius Castoriadis, *Political and Social Writings,* vol. 1, edited by David Ames Curtis (Minneapolis: University of Minnesota Press, 1988), p. 106.
11. I develop this argument further in *The End of Politics: Corporate Power and the Decline of the Public Sphere* (New York: Guilford, 2000). See also Hedges, *Death of the Liberal Class,* ch. 4.
12. See Thomas Frank, *The Wrecking Crew* (New York: Henry Holt and Co., 2008), pp. 157–65, and Robert Kennedy, Jr., *Crimes Against Nature* (New York: Harper Perennial, 2004), chs. 6–8.
13. C. Wright Mills, *The Power Elite* (New York: Oxford University Press, 1956).
14. On the topic of U.S. superpower arrogance, see Robert Jay Lifton, *The Superpower Syndrome* (New York: Nation Books, 2003).
15. Aristotle, *The Politics,* in Michael Curtis, ed., *The Great Political Theories,* vol. 1 (New York: Avon Books, 1981), p. 64.
16. For Spencer's views, see www.Salon.com, July 29, 2011.
17. Naomi Klein, "Capitalism vs. the Climate," *Nation* (November 28, 2011), pp. 11–12.
18. See Eric Hobsbawm, *Primitive Rebels* (New York: W. W. Norton and Co., 1959), ch. 1.
19. On the historical problem of spontaneism and localism in popular movements, see my *The Socialist Tradition: From Crisis to Decline* (New York: Routledge, 1995), pp. 47–56.
20. On the role of nationalism in Communist revolutions, see Chalmers Johnson, *Peasant Nationalism and Communist Power* (Stanford: Stanford University Press, 1962), esp. chs. 1, 6, and 7.
21. Russell Jacoby, *Dialectic of Defeat: The Contours of Western Marxism* (New York: Cambridge University Press, 1981).
22. See David Brock, *The Republican Noise Machine* (New York: Crown Publishers, 2004), introduction and ch. 2.
23. On the historic rise of the West German Greens, see my *Social Movements and Political Power* (Philadelphia: Temple University Press, 1986), ch. 5.
24. Lillian Klutzsch, et al., "What Has Happened to Green Principles in Electoral and Parliamentary Politics?" in Margit Mayer and John Ely, eds., *The German Greens* (Philadelphia: Temple University Press, 1998), pp. 97–127.
25. The motif of an increasing bankruptcy of American politics is probably best set forth by Sheldon S. Wolin, in his *Democracy, Inc.* (Princeton: Princeton University Press, 2008), especially chs. 9 and 10.
26. See Robert Michels' classic *Political Parties* (New Brunswick: Transaction Publishers, 1999), ch. 1.
27. On the ascendancy of global corporate power, see David Rothkopf, *Superclass: The Global Power Elite and the World They Are Making* (New York: Farrar, Strauss, and Giroux, 2008), chs. 1 and 4.

28. James Hansen, *The Storms of My Grandchildren* (New York: Bloomsbury, 2009), ch. 8.
29. Richard Heinberg, *Power Down* (Gabriola Island, BC: New Society Publishers, 2004), ch. 4.
30. On Republican efforts to dismantle large parts of the public sector, see Frank, *The Wrecking Crew.*
31. See Hedges, *Death of the Liberal Class,* chs. 2 and 3.
32. See Michael Parenti, *Contrary Notions* (San Francisco: City Lights Books, 2007), p. 96.
33. Ibid., p. 94.
34. Klein, "Capitalism vs. the Climate," 14.
35. Murray Bookchin was one of the first theorists to argue strenuously for a full-scale reordering of the relationship between human beings and nature. See, among other works, *The Ecology of Freedom* (Palo Alto: Cheshire Books, 1982), ch. 12.

Chapter 2

1. Senator Bernie Sanders' interview on radio station KPFK (Los Angeles), September 19, 2010.
2. Ibid.
3. On escalating corporate profits, see
4. Bill McKibben, *Earth: Making a Life on a Tough New Planet* (New York: Henry Holt and Co., 2010), pp. 75–76.
5. Ibid., p. 38.
6. James Howard Kunstler, *The Long Emergency* (New York: Grove Press, 2006), p. 4.
7. Ibid., pp. 24–25.
8. See Al Gore, "Climate of Denial," *Rolling Stone* (July 7–21, 2011).
9. Lester Brown, *Plan B 4.0* (New York: W. W. Norton, 2009), p. 17.
10. Ibid., p. 57.
11. Ibid., p. 60.
12. Ibid., p. 71.
13. James Hansen, *Storms of My Grandchildren* (New York: Bloomsbury USA, 2009), p. 147.
14. Ibid., p. 250.
15. See IPCC Fourth Assessment Report (2007), in www.enwikipedia.org/wiki/Int-Panel-on-Climate-Change
16. See Hansen, *Storms,* pp. 140–42.
17. Ibid., pp. 142, 160.
18. Ibid., p. 171.
19. On the "Venus Express," see ibid., ch. 10.
20. McKibben, *Eaarth,* pp. 75–76.
21. Kunstler, *Long Emergency,* p. 31.

22. Paul Roberts, *The End of Food* (Boston: Mariner Books, 2008), p. xix.
23. Ibid., pp. 211–12.
24. Ibid., p. 211.
25. "Is Meat Sustainable?", *Worldwatch* (July-August, 2004), p. 12.
26. John Robbins, *The Food Revolution* (Berkeley: Conari Press, 2001), p. 248.
27. Ken Midkiff, *The Meat You Eat* (New York: St. Martin's, 2004), p. 39.
28. Upton Sinclair, *The Jungle* (Tucson: Sharp Press, 2003).
29. For a more extensive discussion of the Humane Slaughter Act, see Robbins, *The Food Revolution*, pp. 211–12.
30. Ibid., pp. 220–22.
31. Michael Pollan, *The Omnivore's Dilemma* (New York: Penguin Books, 2006), p. 318.
32. Robbins, *Food Revolution*, p. 248.
33. Ibid., pp. 291–92.
34. David Pimentel and Marcia Pimentel, "World Population, Food, Natural Resources, and Survival," in Ervin Laszlo and Peter Seidel, eds., *Global Survival* (New York: SelectBooks, 2006), pp. 31–53.
35. Ibid., p. 46.
36. Brown, *Plan B 4.0*, pp. 66–67.
37. On the fast-food industry reshaping of American society, see Eric Schlosser, *Fast Food Nation* (Boston: Houghton-Mifflin, 2001), especially pp. 1–10.
38. Ibid., p. 3.
39. On the beef culture and the Enlightenment, see Jeremy Rifkin, *Beyond Beef* (New York: Penguin Books, 1992), chs. 11–16.
40. Ibid., p. 257.
41. On the Fordist dimension of the fast-food industry, see George Ritzer, *The McDonaldization of Society* (Thousand Oaks: Pine Forge Press, 2000), ch. 2.
42. See Schlosser, *Fast Food Nation*, pp. 149–50; 154; 195.
43. Ritzer, *McDonaldization*, pp. 1–20.
44. Robbins, *Food Revolution*, pp. 60–65.
45. Carol J. Adams, *The Sexual Politics of Meat* (New York: Continuum, 2004), p. 43.
46. On the failed cancer "war," see especially Samuel S. Epstein, *Cancer-Gate* (Amityville, NY: Baywood, 2005), and Ralph W. Moss, *The Cancer Industry* (Brooklyn: Equinox, 1996).
47. On the deadly adverse reactions experienced by hundreds of thousands of Americans yearly to pharmaceuticals, see Sidney Wolfe, et al., *Worst Pills, Best Pills* (New York: Pocket Books, 2006), pp. xxi–xxvi. See also John Abramson, *Overdosed America* (New York: HarperCollins, 2004).
48. Already a classic, see T. Colin Campbell, *The China Study* (Dallas: BenBella Books, 2006).

49. Ibid., p. 305.
50. Ibid., p. 317.
51. Ibid., p. 312.
52. For an overview of American foreign and military policy, see Carl Boggs, *Imperial Delusions* (Lanham, MD: Rowman and Littlefield, 2006), especially introduction and chapter one.
53. Paul Roberts, *The End of Oil* (Boston: Houghton-Mifflin, 2004), p. 12.
54. On the depletion of global oil reserves, see Richard Heinberg, *Power Down* (Gabriola Island, BC: New Society Publishers, 2004), pp. 23–37.
55. On the severe limits of natural gas, both geologically and politically, see ibid., pp. 49–52.
56. Ibid., p. 54.
57. Michael Klare, *Resource Wars* (New York: Henry Holt and Co., 2001), p. 29.
58. See Antonia Juhazs, *The Bush Agenda* (New York: Regan Books, 2006).
59. Ibid., p. 298.
60. John Bellamy Foster, et al., "The U.S. Imperial Triangle and Military Spending," *Monthly Review* (October 2008), p. 14.

Chapter 3

1. Bill McKibbern, *Eaarth* (New York: Henry Holt and Co., 2010), p. 101.
2. C. Wright Mills, *The Power Elite* (New York: Oxford University Press, 1956).
3. Karl Marx, "Manifesto of the Communist Party," in Robert C. Tucker, ed., *The Marx-Engels Reader* (New York: W. W. Norton, 1978), pp. 476–77.
4. Ibid., p. 476.
5. Mills, *The Power Elite*, p. 23.
6. Ibid., p. 12.
7. David Rothkopf, *Superclass* (New York: Farrar, Strauss, and Giroux, 2008), pp. 300–01.
8. Ibid., p. 126.
9. Ibid., p. 143.
10. Arianna Huffington, *Third-World America* (New York: Crown Publishers, 2010), p. 149.
11. G. William Domhoff, *Who Rules America?* (New York: McGraw-Hill, 2006), pp. 158–60.
12. Thomas Frank, *The Wrecking Crew* (New York: Henry Holt and Co., 2008), p. 115.

13. Ibid., p. 95.
14. See Matt Taibbi, "Wall Street's Bailout Hustle," *Rolling Stone* (March 4, 2010).
15. Robert Kennedy, Jr., *Crimes Against Nature* (New York: Harper/Collins, 2004), p. 24.
16. Ibid., p. 34.
17. Ibid., p. 39.
18. Ibid., p. 99.
19. Helen Caldicott, *The New Nuclear Danger* (New York: New Press, 2002), p. 26.
20. On the business structure of Big Pharma, see Donald L. Bartlett and James B. Steele, *Critical Condition* (New York: Doubleday, 2004), ch. 2.
21. See Marcia Angell, *The Truth About the Drug Companies* (New York: Random House, 2004), pp. 214–16.
22. Sidney Wolfe, et al., *Worst Pills, Best Pills* (New York: Simon and Schuster, 2006), p. 6.
23. *Los Angeles Times* (March 9, 2010).
24. Ibid.
25. Quoted in Peter H. Stone, *Casino Jack* (Brooklyn: Melville House, 2010), p. 179.
26. See Neal Gabler, op-ed piece, *Los Angeles Times* (December 6, 2010).
27. See Johann Hari, "The Wrong Kind of Green," *Nation* (March 22, 2010).
28. Ibid., p. 13.
29. See the *Los Angeles Times* (November 19, 2011).
30. *Los Angeles Times* (February 16, 2010).
31. See Kevin Drum, "Capital City," *Mother Jones* (January–February, 2010), p. 50.
32. Robert Auerbach, *Deception and Abuse at the Fed* (Austin: University of Texas Press, 2009).
33. *Time* (December 30, 2009), p. 48.
34. Ibid.
35. Robert Scheer, *The Great American Stickup* (New York: Nation Books, 2010).
36. *Los Angeles Times* (January 28, 2011).
37. Frederick Kaufman, "The Food Bubble," *Harper's* (July 2010), p. 34.
38. Andrew Bacevich, *Washington Rules* (New York: Henry Holt and Co., 2010), p. 27.
39. Peter Irons, *War Powers* (New York: Henry Holt and Co., 2005), p. 272.
40. Ibid., p. 2.
41. Chalmers Johnson, *Nemesis: The Last Days of the American Republic* (New York: Henry Holt and Co., 2007), p. 259.

42. Charlie Savage, *Takeover: The Return of the Imperial Presidency* (New York: Little, Brown, and Co., 2007), p. 75.
43. Johnson, *Nemesis*, p. 95.
44. Ibid., p. 91.
45. James Bamford, *The Shadow Factory* (New York: Doubleday, 2008), p. 304.
46. Ibid., p. 110.
47. Bamford, *The Shadow Factory*, chs 4–7.
48. *Los Angeles Times* (January 30, 2011).
49. Tariq Ali, *The Obama Syndrome* (London: Verso, 2010), pp. 56–57.
50. Robert McChesney, *Rich Media, Poor Democracy* (New York: The New Press, 1999), p. 281.
51. Ben Bagdikian's volume is *The Media Monopoly* (Boston: Beacon Press, 1992).
52. David Brock, *The Republican Noise Machine* (New York: Crown Books, 2004), p. 50.
53. See Peter Phillips and Mickey Huff, eds., *Censored 2010* (New York: Seven Stories, 2009) and *Censored 2011* (same press, 2010).
54. Norman Solomon, *War Made Easy* (Hoboken, NJ: John Wiley and Sons, 2005), pp. 110–11.
55. Sut Jhally, "Advertising at the Edge of the Apocalypse," in Paula Rothenberg, ed., *Race, Class, and Gender in the United States* (New York: Worth, 2010), p. 621.
56. Ibid., p. 627.
57. See Robert McChesney, *The Problem of the Media* (New York: Monthly Review, 2004), p. 138.
58. Sebastian Jones, "The Media-Lobbying Complex," *Nation* (March 1, 2010), pp. 11, 13.
59. See Sheldon S. Wolin, *Democracy, Inc.* (Princeton, NJ: Princeton University Press, 2009), especially chs. 10, 11.
60. Ibid., p. 160.
61. John Nichols and Robert McChesney, "The Money and Media Election Complex," *Nation* (November 29, 2010).
62. Ibid., p. 13.
63. On this point, see the excellent critique by Evgeny Morozov in *The Net Delusion*.
64. See Domhoff, *Who Rules America?*, pp. 154–55.
65. www.OpenSecrets.com.
66. Ibid.
67. *Los Angeles Times* (November 4, 2010).
68. See Ari Berman, "The GOP War on Voting," *Rolling Stone* (September 15, 2011).
69. Chris Hedges, *Death of the Liberal Class* (New York: Nation Books, 2010), pp. 9–10.
70. Ibid., p. 12.

71. Ali, *The Obama Syndrome*, p. 76.
72. Ibid., p. 33.
73. See the *Los Angeles Times* (February 1, 2012).
74. See Zygmunt Bauman, *In Search of Politics* (Stanford: Stanford University Press, 1999), pp. 120–22.
75. Hedges, *Death of the Liberal Class*, pp. 194–95.

Chapter 4

1. Al Gore, *Our Choice: A Plan to Solve the Climate Crisis* (New York: Melcher Media, 2009).
2. Ibid., p. 388.
3. Ibid., p. 372.
4. Ibid., pp. 342–43.
5. Ibid., p. 346.
6. Ibid., p. 320.
7. Ibid., pp. 399–401.
8. Thomas L. Friedman, *Hot, Flat, and Crowded* (New York: Farrar, Straus, and Giroux, 2009), pp. 458–59.
9. Ibid., pp. 212–13.
10. Ibid., pp. 226–27.
11. Ibid., pp. 277–78.
12. Ibid., p. 291.
13. Ibid., p. 298.
14. Ibid., pp. 442, 446.
15. Ibid., p. 455.
16. Lester R. Brown, *Plan B 4.0: Mobilizing to Save Civilization* (New York: W. W. Norton and Co., 2009), p. 242.
17. Ibid., ch. 10.
18. Ibid., pp. 264–65.
19. Ibid., p. 71.
20. Ibid., p. 145.
21. Ibid., p. 237.
22. Ibid., p. 233.
23. Friedman, *Hot, Flat, and Crowded*, p. 291.
24. See Matthew E. Kahn, *Climatopolis* (New York: Basic Books, 2010).
25. *Los Angeles Times* (May 31, 2011).
26. Chris Hedges, *Death of the Liberal Class* (New York: Nation Books, 2010), p. 153.
27. Immanuel Wallerstein, *The End of the World as We Know It* (Minneapolis: University of Minnesota Press, 1999), p. 83.
28. Joel Kovel, *The Enemy of Nature* (London: Zed Books, 2002), pp. 80–81.
29. Max Horkheimer and Theodor W. Adorno, *Dialectic of Enlightenment* (New York: Continuum, 1995), p. 14.

Notes

30. Ibid., p. 24.
31. Friedman, *Hot, Flat, and Crowded*, p. 213.
32. Ibid., p. 217.
33. Ibid., pp. 226–27.
34. Ibid., pp. 277–78.
35. Ibid., p. 305.
36. Gore, *Our Choice*, pp. 399–401.
37. Ibid., pp. 372, 388.
38. Brown, *Plan B 4.0*, p. 243.
39. Ibid., p. 261.
40. Ibid., pp. 265–66.
41. Morley Winograd and Michael D. Hais, *Millenial Makeover* (New Brunswick: Rutgers University Press, 2009), p. 140.
42. Ibid., p. 192.
43. *Los Angeles Times* (February 19, 2011).
44. Richard Heinberg, *Power Down* (Gabriola Island, BC: New Society Publishers, 2004), p. 132.
45. Michael Ruppert, *Confronting Collapse* (White River Junction, VT: Chelsea Green Publishing, 2009), p. 98.
46. Ibid., pp. 98–117.
47. James Howard Kunstler, *The Long Emergency* (New York: Grove Press, 2006), p. 138.
48. Heinberg, *Power Down*, p. 120.
49. Kunstler, *Long Emergency*, p. 134.
50. Cited in Ruppert, *Confronting Collapse*, p. 122.
51. Ibid., p. 120.
52. Helen Caldicott, *Nuclear Power Is Not the Answer* (New York: New Press, 2006), p. viii.
53. Kunstler, *Long Emergency*, pp. 145–46.
54. A summary of the California report on solar energy is contained in the *Los Angeles Times* (November 9, 2011).
55. Ibid., p. 17.
56. For an overview of Green politics, see John Ely, "Green Politics in Europe and the United States," in Margit Mayer and John Ely, eds., *The German Greens* (Philadelphia: Temple University Press, 1998), pp. 193–209.
57. Hedges, *Death of the Liberal Class*, p. 153.

Chapter 5

1. See Johann Hari, "The Wrong Kind of Green," *Nation* (March 22, 2010), p. 13.
2. Ibid., p. 18.
3. Richard Heinberg, *Power Down* (Gabriola Island, BC: New Society Publishers, 2004), p. 54.

4. On the "consensus trance" and "sleepwalking into the future," see James Howard Kunstler, *The Long Emergency* (New York: Grove Press, 2005), pp. 1–21.
5. Bill McKibben, *Earth* (New York: Henry Holt and Co., 2010), p. 102.
6. Kunstler, *Long Emergency*, p. 65.
7. McKibben, *Eaarth*, ch. 4.
8. Kunstler, *Long Emergency*, p. 286.
9. Murray Bookchin, *The Philosophy of Social Ecology* (Montreal: Black Rose Books, 1990), pp. 115–16.
10. Murray Bookchin, *Remaking Society* (Montreal: Black Rose Books, 1989), p. 15.
11. Murray Bookchin, *The Modern Crisis* (Philadelphia: New Society Publishers, 1986), p. 99.
12. Ibid., p. 106.
13. Ibid., p. 121.
14. Ibid., p. 50.
15. Ibid., pp. 52–53.
16. Ibid., p. 59.
17. Ibid., p. 67.
18. Ibid., p. 182.
19. Ibid., pp. 184–85.
20. Murray Bookchin, *Re-enchanting Humanity* (London: Cassell, 1995), p. 236.
21. See George Sessions, ed., "Ecocentrism and the Anthopocentric Detour," in *Deep Ecology for the Twenty-First Century* (Boston: Shambala, 1995), pp. 169–77.
22. See, for example, Jack Turner, "In Wilderness is the Preservation of the World," in Sessions, ed., *Deep Ecology*, pp. 331–38.
23. Arne Naess, "Equality, Sameness, and Rights," in Sessions, ed., *Deep Ecology*, p. 222.
24. See Arne Naess, "Deep Ecology and Lifestyle," in Sessions, ed., *Deep Ecology*, p. 260.
25. Arne Naess, "The Deep Ecological Movement," in Sessions, ed., *Deep Ecology*, p. 68.
26. Timothy W. Luke, *Ecocritique* (Minneapolis: University of Minnesota Press, 1997), p. 23.
27. Chris Williams, *Ecology and Socialism* (Chicago: Haymarket Books, 2010), p. 13.
28. Ibid., p. 230.
29. For an excellent discussion of productivism (and related issues) in Marxist theory, see Ronald Aronson, *After Marxism* (New York: Guilford, 1995), pp. 90–123.
30. See, for example, "Ecology, Capitalism, and the Socialization of Nature," an interview with *Monthly Review* editor John Bellamy Foster, in *Monthly Review* (November 2004), pp. 1–12.

31. Ted Benton, *Natural Relations: Ecology, Animal Rights, and Social Justice* (London: Verso, 1998), pp. 23–31.
32. William Leiss, *The Domination of Nature* (New York: George Braziller, 1972), p. ix.
33. For an extensive discussion and critique of the council tradition, see Carl Boggs, *The Two Revolutions: Gramsci and the Dilemmas of Western Marxism* (Boston: South End Press, 1984), pp. 69–118.
34. On the role played by a theorization of ideological hegemony in Western Marxism, see Boggs, *The Two Revolutions*, pp. 153–98.
35. For a comprehensive discussion of the emergence of left Greens in Europe, see Rudolf Bahro, *From Red to Green* (London: Verso, 1984), ch. 8.
36. Immanuel Wallerstein, "Structural Crisis in the World-System," *Monthly Review* (March 2011), p. 35.
37. Ibid., p. 39.
38. John Bellamy Foster, "Ecology and the Transition from Capitalism to Socialism," *Monthly Review* (November 2008), p. 8.
39. Williams, *Ecology and Socialism*, pp. 233, 235.
40. Bahro, *From Red to Green*, p. 219.
41. See Rudolf Bahro, "The SPD and the Peace Movement," *New Left Review* (January-February, 1982), p. 46.
42. See in particular Gene Sharp's *Making Europe Unconquerable* (London: Taylor and Francis, 1985).
43. Petra Kelly, *Fighting for Hope* (Boston: South End Press, 1984), pp. 66–68.
44. Jon Burchell, *The Evolution of Green Politics: Development and Change Within European Green Parties* (London: Earthscan, 2002).

Chapter 6

1. Joel Kovel, *The Enemy of Nature* (London: Zed Books, 2002), p. 21.
2. See, for example, John Ely's prescient analysis in "Green Politics in Europe and the United States," in Margit Ely and John Mayer, eds., *The German Greens* (Philadelphia: Temple University Press, 1998), pp. 193–209.
3. For a general treatment of the relationship between social-movement trajectories and the requirements of political strategy, see Carl Boggs, *Social Movements and Political Power* (Philadelphia: Temple University Press, 1986), ch. 6.
4. On the spontaneist character of the American New Left, see especially Judith Clavir Albert and Stewart Edward Albert, eds., *The Sixties Papers* (Westport, CT: Praeger, 1984), pp. 10–63.
5. On Gramsci's concept "social bloc" and its larger historical context, see Boggs, *The Two Revolutions: Gramsci and the Dilemmas of Western Marxism* (Boston: South End Press, 1984), pp. 282–90.

6. On the historic emergence of what is often called "transnational counterpublics," see Francis Shor, *Dying Empire* (New York: Routledge, 2010), pp. 210–15.
7. Seymour Melman, *The Demilitarized Society* (Montreal: Harvest House, 1988), p. 52.
8. On the severe limits of liberal environmentalism, see Chris Hedges, *Death of the Liberal Class* (New York: Nation Books, 2010), pp. 184–95.
9. See Charles Reich, *The Greening of America* (New York: Random House, 1971).
10. Paul Gilding, *The Great Disruption* (New York: Bloomsbury, 2011), p. 2.
11. Ibid., p. 263.
12. Frances Moore Lappe, *Eco Mind* (New York: Nation Books, 2011), p. 10.
13. Ibid., p. 189.
14. Ibid., p. 194.
15. *Los Angeles Times* (January 31, 2012).
16. Lappe, *Eco Mind*, p. 21.
17. Robert Michels, *Political Parties* (New Brunswick: Transaction Publishers, 1999), pp. 238–53.
18. On the theory of deradicalization as applied to the early European social-democratic experience, see Peter Gay, *The Dilemma of Democratic Socialism* (New York: Collier Books, 1962); Michels, *Political Parties;* Guenther Roth, *The Social Democrats in Imperial Germany* (Totowa, NJ: Bedminster Press, 1963); and Carl Schorske, *German Social Democracy* (New York: John Wiley and Sons, 1955). See also Carl Boggs, *The Socialist Tradition* (New York: Routledge, 1995), chs. 1, 2, and 5.
19. On the rise and decline of European Communist parties, see Boggs, *The Socialist Tradition*, ch. 4.
20. For an overview of developments helping to give rise to Western Marxism, see Roger Gottlieb, *Marxism: Origins, Betrayal, Rebirth* (New York: Routledge, 1992), ch. 5.
21. For a general overview and critical assessment of Gramci's concept of ideological hegemony, see Boggs, *The Two Revolutions*, ch. 5.
22. Gramsci here refers to "new initiatives" and "cathartic moments" of popular resistance that challenge the structures of domination, which no longer appear as "external forces" that crush the human spirit. See Gramsci, "The Study of Philosophy," in *Selections from the Prison Notebooks* (New York: International Publishers, 1971, pp. 366–67.
23. See Chalmers Johnson, *Peasant Nationalism and Communist Power* (Stanford: Stanford University Press, 1962), on the binding connection between nationalism and revolution.
24. See Gramsci, "Americanism and Fordism," in *SPN*, pp. 285–87.

25. On the general impact of capitalist rationalization on American society, see Max Horkheimer and Theodor W. Adorno, *Dialectic of Enlightenment* (New York: Continuum, 1995); Harry Braverman, *Labor and Monopoly Capital* (New York: Monthly Review Press, 1974); Herbert Marcuse, *One-Dimensional Man* (Boston: Beacon Press, 1964), and Edwards, *Contested Terrain* (New York: Basic Books, 1979).
26. Marcuse, *One-Dimensional Man*, p. 159.
27. Ibid., p. 245.
28. Ibid., p. 259.
29. Herbert Marcuse, *Counterrevolution and Revolt* (Boston: Beacon Press, 1972), p. 61.
30. Ibid., p. 74.
31. Ibid., pp. 61–62.
32. Murray Bookchin, *Remaking Society* (Montreal: Black Rose Books, 1989), p. 169.
33. Ibid., p. 172.

Conclusion: A Green Politics?

1. See Tom Mertes, *A Movement of Movements* (London: Verso, 2004), p. 244.
2. Peter McLaren, *Capitalists and Conquerors* (Lanham, Md.: Rowman and Littlefield, 2009), p. 173.
3. See David Orr, "The Ecological Deficit: Creating a New Political Framework", in Richard Heinberg and Daniel Lerch, eds, *The Post-Carbon Reader* (Healdsburg, CA.; Watershed Media, 2010), p. 65.
4. Ibid., p. 70.
5. jJohn Nichols, The "*S*" *Word* (London: Verso, 2011), p. 258.
6. John Bellamy Foster, "The Ecology of Marxian Political Economy", *Monthly Review* (September, 2011), p. 14.
7. Chris Williams, *Ecology and Socialism* (Chicago: Haymarket Books, 2010), p. 233.
8. Joel Kovel, *The Enemy of Nature* (London: Zed Books, 2002), p. 226.
9. Ibid., pp. 233–34.

Postscript: Ecology and Population

1. Chris Hedges, *The World as It Is* (New York: Nation Books, 2010), p. 271.
2. *New York Times* (May 2, 2011).
3. Chris Williams, *Ecology and Socialism* (Chicago: Haymarket Books, 2010), pp. 33, 66.
4. Ibid., pp. 76–68.
5. Ibid., p. 49.

6. See the *Los Angeles Times* (June 22, 2011).
7. Paul R. Ehrlich and Anne H. Ehrlich, "The Population Explosion," www.jayhanson.us/page 27.html (p. 1).
8. Ibid., p. 2.
9. Ibid.
10. Williams, *Ecology and Socialism*, p. 42.
11. Bill McKibben, *Eaarth* (New York: Henry Holt and Co., 2010), p. 44.
12. Sandra Postel, "Water: Adapting to a New Normal," in Richard Heinberg and Daniel Lerch, eds., *The Post Carbon Reader* (Healdsburg, CA: Watershed Media, 2010), p. 80.
13. Ibid., p. 87.
14. Lester Brown, *Plan B 4.0* (New York: W. W. Norton and Co., 2009), pp. 226–27.
15. John Robbins, *The Food Revolution* (Berkeley: Conari Press, 2001), pp. 236–37.
16. On the dramatic global ramifications of the American fast-food industry, see Eric Schlosser, *Fast-Food Nation* (Boston: Houghton-Mifflin, 2001), ch. 10.
17. Brown, *Plan B 4.0*, p. 234.
18. Williams, *Ecology and Socialism*, p. 47.
19. Brown, *Plan B 4.0*, p. 233.
20. David and Marcia Pimentel, "World Population, Food, Natural Resources, and Survival," in Ervin Laszlo and Peter Seidel, eds., *Global Survival* (New York: SelectBooks, 2006), p. 45.

Index

Abramoff, Jack, 67
Adams, Carol, 43
Adelson, Sheldon, 95
Adorno, Theodor, 108, 188
Ali, Tariq, 6, 81, 94
Alliance for Climate Protection, 69
American Legislative Effectiveness Committee, 69
Animal Liberation Front, 161
Atkins, Robert, 41
Auerbach, Robert, 73

Bagdikian, Ben, 82
Bahro, Rudolf, 143, 145, 146, 190
Bamford, James, 80
Bartlett, Donald, 66
Benton, Ted, 133
Berkeley Earth Surface Temperature Project, 32
Bernanke, Ben, 73
Bernstein, Eduard, 135, 175, 178, 183
Big Pharma, 65, 66
Bookchin, Murray, 125–8, 139, 190
BP Corporation, 68
Braverman, Harry, 188
Brazil, 203, 205
Brock, David, 14, 83
Brown, Lester, 3, 29, 101–5, 110, 112, 113, 118, 197, 200
Bullitt Center (Seattle), 111
Bush, George-II Presidency, 51, 77, 78, 83

Caldicott, Helen, 65, 116
Campbell, Colin, 43, 44

Capitalism, xvi, 118, 132, 133, 138, 174, 176, 184, 202
  and State Capitalism, 187
  and World Capitalist System, 2
Carbon Dioxide Analysis Center, 2
Carson, Rachel, 125
Central Intelligence Agency, 78, 79, 80
Chamber of Commerce (U.S.), 67, 68, 70, 90
China, 5, 33, 52, 55, 172, 203–6
Chopra, Deepok, 162, 163
Civil Rights movement, 180–2
Clean Air Act, 19, 68, 99
Clear Channel Corporation, 83
Climate-Change Conference (Durban), 3
Clinton, Bill Presidency, 19
Cohn-Bendit, Daniel, 143
Comcast Corporation, 86
Commoner, Barry, 125
Communism, 8, 136, 138, 155, 167, 176, 186, 192
ConAgra Plant, 41
Congress Party (India), 181
Consumer Protection Finance Agency, 71
Copenhagen World Summit, 10, 11, 166
Corporate growth machine, xvi
Crossroads GPS, 91, 95

Deep Ecology, 129–31, 141, 151
DeMint, Jim (Sen), 91
Democrats, 90, 93, 107, 116, 165
Deradicalization, 174
Domhoff, G. William, 85
"Dual Power", 176

Earth First!, 161
Earth Policy Institute, 101
Ecosocialism, 131, 139, 154, 193, 194
Ehrlich, Paul and Anne, 197, 200, 201
England, 181
Enlightenment, 132–5, 153, 183, 187
Environmental Action, 160
Environmental Defense Fund, 70, 161
Environmental Protection Agency, 9, 19, 25, 63, 64, 69
European Parliament, 168
European Union, 166
Exxon Mobil Corporation, 68

Fanon, Franz, 179
Federal Bureau of Investigation, 80, 81
Federal Communications Commission, 9, 63
Federal Humane Slaughter Act, 38
Federal Reserve Bank, 72, 73
Financial Services Roundtable, 71
Food and Drug Administration, 63, 66
"Fordism" (Gramsci), 187, 189
Foreign Intelligence Surveillance Act, 80
Foster, John Bellamy, 131, 140, 141
FOX News Network, 86, 87
Frankfurt School, 184
Frank, Thomas, 7, 63
Friedman, Thomas L., 100, 101, 103–5, 108–10, 118
Fundis (Greens), 146

Gandhi, Mahatma, 178, 179, 181, 182
Geithner, Timothy, 73, 74
Gilding, Paul, 162
Gingrich, Newt, 95
Global Coherence Project, 163

Global Exchange, 160
Globalization, 107
Global Justice Ecology Project, 160
Goldman Sachs, 74
Gore, Al, 16, 98, 99, 103–5, 109, 110, 112, 118
Gramsci, Antonio, 81, 82, 136, 138, 156, 184–9
Green Festivals, 160
Greenpeace, 161
Greens, xvii, 14, 15, 119, 142–9, 177, 182, 193
    and American Greens, 16, 148–50, 157
    and European Greens, 119, 149, 153, 155, 157, 168, 180, 191
    and Left Greens, 128, 190
    and West German Greens, 142, 145–7, 159, 174, 175
Gross Domestic Product, 167–74
Guevara, Che, 179

Hais, Michael, 110
Hansen, James, 17, 31, 32, 33
Hari, Johann, 121, 122
Hedges, Chris, 6, 93, 95, 120, 202, 206
Heinberg, Richard, 47, 115, 122–4
Hobsbawm, Eric, 12
Horkheimer, Max, 198
Huffington, Arriana, 62
Humane Society, 161

IBP Corporation, 38
India, 5, 36, 54, 181, 200, 203, 204
Institute for Social Ecology, 160
International Criminal Court, 5, 32
International Panel on Climate Control, 2, 17
Iran, 52–6
Iraq Liberation Act, 49
Irons, Peter, 66

Israel, 52, 53
 and Israel Lobby, 54
Italy, 186

Jacobinism, 21, 154, 174
Jacoby, Russell, 14
Jhally, Sut, 84
Johnson, Chalmers, 78, 79

Kahn, Matthew, 105
Kautsky, Karl, 132
Kelly, Petra, 143, 145
Kennedy, Robert, Jr, 7, 64
King, Martin Luther, 168
Klare, Michael, 4, 48
Klein, Naomi, 9, 19, 21
Koch Brothers, 64, 69, 91, 92, 93
Kovel, Joel, 153, 194, 195, 197
Kunstler, James Howard, 26, 27, 115, 122–4

Lappe, Frances Moore, 160, 163
Leiss, William, 133
Leninism, xvii, 13, 118, 135, 136, 141, 143, 155, 177, 178, 194
Lenin, V. I., 183
Liberalism, 106, 117, 118, 126, 161, 183
Liberal tradition, 97
Luke, Tim, 131
Luxemburg, Rosa, 6, 136, 137

Malthus, Robert/Malthusian, 107–200, 201, 202, 206
Manhattan Project, 112
Marcuse, Herbert, 125, 138, 187–90
Marxism, 23, 24, 126, 131, 132, 134, 136, 138–42, 154, 165, 175, 182, 183, 188, 193
 and "Western" Marxism, 138, 184
Marx, Karl, 59, 60, 132–4, 140, 198
McChesney, Robert, –88, 82, 85
McDonalidization, 41

McKibben, Bill, 25, 26, 33, 57, 122, 124, 203
McLaren, Peter, 192–4
Melman, Seymour, 158
Mexico, 205
Michels, Robert, 16
Mills, C. Wright, 8, 58, 59, 60, 89, 187
MoveOn.org, 163–5
Murdoch, Rupert, 83, 86

Nader, Ralph, 148
National Climate Data Center, 30
National Oceanic and Atmospheric Administration, 2
National Security Agency, 79, 80
National Security Operations Center, 80
National Security State, 50, 76, 77, 79, 81
NATO, 147
New Deal, 92, 120
New Left, 156
Nichols, John, 87, 88, 193
Nigeria, 198
Nuclear Non-Proliferation Treaty, 5, 53, 54
Nuclear Regulatory Agency, 65

Obama, Barack Presidency, 6, 18, 28, 69, 71, 81, 90, 94, 110, 111
Occupational Safety and Health Administration, 25
Occupy Movement, 7, 20, 148, 159, 164

Parenti, Michael, 21
Pentagon, 18, 25, 33, 51, 52, 76, 171
Pimentel, David and Marcia, 40, 206
Planet Drum Foundation, 160
Pollan, Michael, 38
Populists, 159
Post-Carbon Institute, 192, 193
Postel, Sandra, 203

Rand, Ayn, 92
Reagan, Ronald, 6, 176
Realos (Greens), 146, 174, 195
Reich, Charles, 162
Republicans, 9, 67, 69, 74, 75, 159
Rifkin, Jeremy, 40
Robbins, John, 38
Roberts, Paul, 35, 36, 46
Romney, Mitt, 94
Roosevelt, Theodore
    Administration, 77
Rothkopf, David, 61
Ruppert, Michael, 114, 122–4
Russia, 203
Ryan, Paul (Rep.), 92

Sanders, Bernie, 25
Savage, Charlie, 78
Scheer, Robert, 73
Schlosser, Eric, 40
Schumer, Charles (Sen.), 72
Sea Shepherd Society, 161
Second International, 166, 175
Sessions, George, 129
Sharp, Gene, 145
Sierra Club, 70, 161
Simmons, Harold, 95
Sinclair, Upton, 38
Social Ecology, 125–30
Socialism, 176, 183
Socialist International, 166, 167
Solomon, Norman, 84
Somalia, 205
Sorel, Georges, 179
Soviet model, 135, 192

Spencer, Roy, 9
Stalin, Josef, 166
Steele, James, 66

Tea Party, 19, 20, 63, 86, 91, 92, 93, 120
Tennessee Valley Authority, 9, 64, 115
Thatcher, Margaret, 176
Think City (Indiana), 117
Third International, 135
Togliatti, Palmiro, 168
Trotsky, Leon, 136, 166, 167
    and Fourth International, 167
Truman, Harry Administration, 76

Wallerstein, Immanuel, 107, 139, 140
Wall Street, 4, 20, 54, 70, 71, 72, 75, 159
Wal-Mart, 106, 188
Warfare State, 76
War Powers Act, 77
Weber, Max, 59, 60, 187, 188
Williams, Chris, 131, 141, 194, 198–200, 201, 202, 204–6
Winograd, Morley, 110
Wolfe, Sidney, 66
World Bank, 6, 39, 61, 166
World Social Forum, 167
World War I, 178
World War II, 45, 100, 186
World Wildlife Fund, 70

Zapatistas, 177